# GUIDE TO
# SEASHORES
## AND SHALLOW SEAS
### OF BRITAIN AND NORTHERN EUROPE

ANDREW CAMPBELL

First published in 2005 by Philip's,
a division of Octopus Publishing Group Ltd,
2–4 Heron Quays, London E14 4JP

ISBN-13 978-0-540-08747-1
ISBN-10 0-540-08747-5

A CIP catalogue record for this book is available
from the British Library.

Printed in China

COMMISSIONING EDITOR Frances Button
EDITOR Joanna Potts
EXECUTIVE ART EDITOR Mike Brown
DESIGNERS Caroline Ohara, Will Butler
PRODUCTION Sally Banner

### Other titles from Philip's

Philip's publish a range of titles
in the same format as this book,
including:

*Philip's Guide to Weather*
*Philip's Guide to Gems*
*Philip's Guide to Fossils*
*Philip's Minerals, Rocks and Fossils*
*Philip's Guide to Seashells of the World*
*Philip's Guide to the Oceans*

### For details of Philip's products

website: www.philips-maps.co.uk
e-mail: philips@philips-maps.co.uk
tel: 020 7644 6940
fax: 020 7644 6986

FRONT COVER: clockwise from top left *Rhizostoma pulmo*,
*Asterina phylactica*, *Aequipecten opercularis*, *Labrus
mixtus*, *Strongylocentrotus droebachiensis*, *Hippocampus
guttulatus*, *Urticina eques*, *Fucus spiralis*, *Cassidaria
echinophora.*
Centre *Cancer pagurus*.

BACK COVER: Left *Lomentaria clavellosa*, Right *Halcampa
chrysanthellum*

### Preface

This book is intended to provide the
layman and the student with a simple
means of identifying most of the common
marine plants and animals in the field.
The accurate determination of certain
species often depends on the correct use
of an identification key in a specialist's
monograph. Unfortunately the use of
such keys frequently presupposes some
knowledge of the group of organisms
in question and may not therefore be
of great help to the beginner. As an
alternative, this book provides illustrations and simple accounts of the form,
habitats and distribution of a number of
species to be found on the shores and in
the shallow seas around Britain and
Europe. It must be said, however, that a
book of this size cannot always provide
an exact identification, especially where
some of the more obscure groups are
concerned. For this reason the reader is
directed to other sources at certain
points in the text.

Many thousands of species have been
recorded in the European seas and it
would be impossible to include them all
in this book. The thousand or so selected
for inclusion have been chosen because
they are relatively common, and because
they can generally be identified without
the aid of a microscope. In a number of
cases the use of a hand lens is called for.

**Andrew Campbell**

# CONTENTS

# INTRODUCTION

For generations, people living by the shores of Europe have looked to the sea as a source of food and a means of trading, and consequently have become aware of the diversity of plants and animals existing there. For the biologist the seashore is an excellent training ground; for the layman it is a fascinating area for exploration. Although Aristotle was one of the first to describe marine organisms, our knowledge of them developed slowly. Today it is considerable, and almost entirely due to work undertaken in the last two hundred and fifty years. In 1758 the Swedish Botanist, Linnaeus, published *Systema Naturae* and established a method for classifying organisms which survives today. Later, in 1788, O. F. Müller described and illustrated marine animals from Scandinavia for the first time. A number of devoted and energetic naturalists emerged in the nineteenth century, and in almost every European maritime nation they collected, described and classified the easily obtainable plants and animals. In 1832 two Frenchmen, Audouin and Milne-Edwards, realized that animals and plants do not occur at random on the shore, and they produced the first scheme for zonation. Zonation was discussed further by Forbes in 1841. Later, in 1859, he published *The Natural History of the European Seas* which provided a general account of the marine biology of Europe.

Meanwhile in England P. H. Gosse was writing popular books about seashore life. These greatly influenced his Victorian readers, many of whom were sadly ignorant of conservation. As a result, many beaches were stripped of the rarer species by the ardent collectors he inspired. Four of his books, *A Naturalist's Rambles on the Devonshire Coast*, 1853; *The Aquarium*, 1854; *Tenby*; *A Seaside Holiday*, 1856; and *A Year at the Shore*, 1865 must be mentioned. In 1859 Darwin's *Origin of Species* was published. This had a profound effect on the future course of biological science. It may have led Dr Anton Dohrn among others to study the development and growth of animals. Dohrn required a marine laboratory where live animals could be kept for study and experiment. He ultimately succeeded in setting up such an establishment at Naples in 1874. Although a marine station had been established at Kristineberg in Sweden much earlier in 1830, the Naples laboratory was the first to be specifically designed for experimental purposes. The potential for such work in marine laboratories was quickly realized, and towards the end of the century similar institutions began to appear all over Europe.

The first in Britain was established at St Andrew's University in 1884, although Sir John Murray had already started work in his floating laboratory 'The Ark', which was moored at Granton. Also in 1884 the Marine Biological Association of the United Kingdom was formed to promote scientific research and to increase knowledge of fishes and fisheries. It opened its famous laboratory at Plymouth in 1888. Now over forty such laboratories are run by universities and governments in Europe. Most are engaged in academic or applied research and their influence on the development of marine science has been enormous. Consequently there exists today a wealth of literature in the form of books and scientific papers, some of which are referred to in the bibliography at the end of this book. Four must be mentioned here, however. *The Sea Shore* by Sir Maurice Yonge was first published in 1949. It gives an excellent account of intertidal life, whilst the companion volumes entitled *The Open Sea* (parts I and II) by Sir Alister Hardy tell of life in the seas and oceans. These appeared in 1956 and 1959. In 1964 J. R. Lewis produced an important work *The Ecology of Rocky Shores* which sets out in a modern, scientific fashion the principles of zonation, and which builds on earlier theories such as those established by T. A. and A. Stephenson in 1949. Three valuable recent references on seashore biology are Cremona, J. 1988, Hawkins, S. J. and Jones, H. D. 1992 and Little, C. and Kitching, J. A. 1996. Campbell, A. C. and Dawes, J. 2005 give more information on the general biology of aquatic animals.

## How to use this book

This book describes the more common plants and animals which may be found on the European shores and in the adjacent shallow seas. The use of a hand lens will be helpful in dealing with some of the smaller species, but in general the naked eye should suffice. Many marine plants and animals lack the common names of their terrestrial counterparts, thus the only one that can normally be used is the scientific name, but where a common name exists it has been included in the text and index. Scientific names are written in italics. Sometimes two scientists have independently given an organism different names. Normally one of these takes preference by common consent, but where an alternative name or synonym persists in use, it is given here. To help further, the author who first used the species name has his name written after it. If his species has subsequently been transferred to another genus the author's name is placed in brackets. In the case of many plants the name of the author who made the transfer is usually also given.

If you think you know the name of a plant or animal you have found, it can be checked in the index. If you wish to identify a specimen for the first time, use the outline key on pages 16–19 to establish where in the book the organism you have found is described, then turn to the page(s) mentioned and make your identification. Almost all the species described in the text are illustrated, and key anatomical features have been given in many cases to help distinguish between similar species; these features will not always be consistent from group to group, however. Line drawings are also used in the text in certain cases. Very often the geographical locality or the habitat, for instance *upper shore* or *in mud,* will assist in helping with the correct identification, so the text and the illustrations must be used together. It will be noted that the distribution of organisms in the sea with regard to depth is not given with equal precision in every case. This is because records themselves are variable. In some cases the exact depths at which organisms occur are known; in other cases only a general impression is available, for instance shallow water. Unfortunately, various authorities have interpreted the words shallow and deep differently. In this book shallow is broadly taken to mean down to 10m, while deep means below 10m, but still on the continental shelf and therefore not below 300m.

In addition to the outline shape, many other factors are important when identifying specimens. The illustrations do not themselves indicate size, but a guide to the general proportions of the various species is given in the text. It must be remembered that juveniles are frequently smaller than the average size of adults, however, and occasionally abnormally large specimens may be found. Colours in marine organisms can be surprisingly variable, and although the artist has taken great trouble to produce life-like impressions of colour, a book of this size cannot include all the variations. Some invertebrates – notably the octopus and cuttlefish – as well as certain fishes, can change their colour rapidly at will. Other specimens alter their coloration when removed from the sea even after a short period of time. Where colour changes are likely to cause confusion reference to alternative characters will usually assist with identification. Growth form and behavioural characters such as locomotion patterns also help in some cases. It should be remembered that the appearance of many species around our coasts is related to the seasons. A number of plants are at their peak in Spring and Summer, some even in Winter, and most animals have seasonal breeding cycles. These may involve migration in fishes. Remember also that many marine organisms are nocturnal, so the experienced marine naturalist should note that some organisms may only be found at night.

From time to time the text includes a reference to a work which will be of particular assistance for the group of organisms in question. A more detailed list of reference sources for all groups is given in the bibliography.

Finally, it should be noted that the male and female of some species have a different external appearance. Where a particular sex is illustrated, ♂ denotes male, and ♀ denotes female.

## The European seas

This book covers an area which includes several widely differing marine regions. These range from the Atlantic Ocean bordering the Iberian Peninsula (sometimes known as the Lusitanian region – Lusitania being an ancient name for Portugal), Ireland, Scotland and Norway, as well as the almost enclosed Baltic Sea. Between these extremes of open ocean and enclosed sea lie the North Sea, the Irish Sea and the English Channel, all of which are influenced by ocean and by land. The size of the European continent means that the climatic conditions vary greatly over the whole area, so increasing the diversity of habitats.

The geography of Europe today is quite different from that of earlier geographical times. For millions of years a large tropical sea known as the Tethys separated the northern continents from the southern, linking what are now the Arctic and Atlantic Oceans with the Indo-west Pacific to the east. In the successive upheavals of the Tertiary period 63–65 million years ago which changed the shape of the land masses, the Tethys was reduced. The Mediterranean was formed from an area previously covered by part of it, and certain species which are found there today have descended from Arctic, Atlantic and Indo-west Pacific ancestors which moved into the Tethys first. They have been able to survive there because of the unique conditions which prevail. The neighbouring parts of the Atlantic ocean provide habitats for a number of Mediterranean organisms. Some of these spread northwards with others of Atlantic origin from Gibraltar to the entrance to the English Channel, and are known as Lusitanian species. Beyond this point those which cannot survive without pure ocean water disappear, but so far as the plankton is concerned the Lusitanian effect may spread up the west coast of Ireland and north as far as Shetland and the northern North Sea in favourable years. The central and northern coasts of Britain and Ireland have been kept relatively mild by the warming effects of the North Atlantic Drift (colloquially called the Gulf Stream) which sweeps moderately warm Atlantic water around the north of Scotland. Some of this water extends into the northern North Sea so that the north-east coast of Scotland carries a greater diversity of Atlantic species than can be found further south. Temperature changes resulting from global warming may melt the polar ice and increase freshwater flowing into the Atlantic. This could weaken the effects of the North Atlantic Drift on north-western Europe and affect the climate. Currently a number of fish species, characteristic of warmer water, are being recorded in the English Channel and Southern North Sea. Generally the waters of the North Sea and the English Channel are slightly less saline than those of the Atlantic. Consequently a restricted Atlantic flora and fauna prevail there, but a few Arctic species extend into these regions by virtue of currents from the north, in addition to running down the north-western coasts of Britain to reach Ireland and the Irish Sea. So far as regional distributions are concerned, the term *Atlantic* in the following text includes the entire western seaboard of Europe from Gibraltar to the Shetlands and Norway. It takes in the Bristol Channel and the Irish Sea, which are mentioned separately only when necessary.

The Baltic is surrounded by much colder land masses than the Mediterranean, and the influence of fresh water is far greater. Apart from the Kattegat and the Skagerrak which somewhat resemble the North Sea, the remainder of the Baltic shows peculiar physical conditions. In the north Baltic lies an area of really deep water, and this trench is filled with salt water over which floats less saline water. Here a number of fully marine organisms occur. The further one penetrates northwards the lower the salinity becomes. Consequently fewer marine species can survive here, although some freshwater species which can tolerate brackish conditions live here as well as genuinely brackish water species which cannot survive in either pure sea water or fresh water. Furthermore the Baltic is very cold and in the Gulfs of Bothnia and of Finland it may be frozen over for nearly half the year. Like the Mediterranean, it contains some species that were left behind after the decline of the Tethys Sea, but most of its marine inhabitants originate from the North Sea.

► Map showing the extent of the various regions mentioned in the book, also indicating the nature of the prevailing physical conditions.

37‰ Salinity in parts per 1000

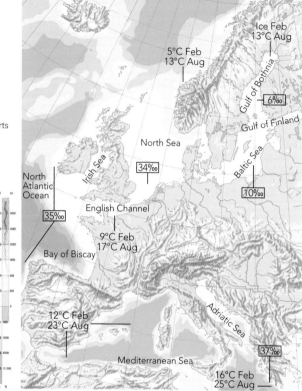

Ice Feb
13°C Aug

5°C Feb
13°C Aug

Gulf of Bothnia

6‰

Gulf of Finland

North Sea

North Atlantic Ocean

Irish Sea

34‰

Baltic Sea

10‰

English Channel

35‰

9°C Feb
17°C Aug

Bay of Biscay

12°C Feb
23°C Aug

Adriatic Sea

37‰

Mediterranean Sea

16°C Feb
25°C Aug

## Life on the seashore

The seashore is that area of the coast which lies between the highest level to which the tides flow and the lowest level to which they ebb. All plant and animal life between these two points is subjected to the movement of the tides and the various side effects which they create. Tidal movement is brought about principally by the gravitational pull of the sun and the moon upon the vast masses of water in the oceans. Variations in ocean depth and the presence and shape of nearby land masses can modify the extent of the tides, and we see these effects particularly in the European area. The tidal range around much of Britain is between 2 and 5m, but in the Baltic Sea and the Mediterranean Sea there is frequently little or no tide, whilst in certain places, for instance the Bristol Channel, tides of 10m or more can occur. Normally the tidal cycle at any one place is repeated just over every 12 hours so that on successive days both the high and low waters appear on average about 50 minutes later than on the day before. In a few areas the form of the coastline is such that an extra tide is inserted between the main tides. Thus in the Solent and at other points around Britain there is an extra high tide tucked in before the low tide period.

Although the twelve-hourly rhythm of the tides is their most conspicuously variable feature, there are others. At new Moon and at full Moon the Sun and the Moon are pulling together almost in a straight line so that the tidal ranges at these times are greater. Such tides are known as *spring tides*. During the intervening periods the pull of the Sun and the Moon are less aligned and may be at right angles, thus reducing their combined effects, so that the tidal ranges are reduced. These are known as *neap tides*. Thus there is a monthly rhythm of spring and neap tides. In addition to this monthly rhythm there is also an annual rhythm, since at the equinoxes (March and September), the Sun and the Moon pull exactly in a straight line giving an exceptionally large tidal range at the spring tides. Conversely in December and June the spring tides are of a smaller range than at any other time of the year. The diagram below shows a diagrammatic representation of the sun and moon's gravitational influence on the tides. The graph on page 10 shows the spring/neap tide cycles for 12 days at a given point on the coast.

The consequences of tidal movements on the shore are considerable. Whilst the greater area will be covered and uncovered by the sea once every 12 hours, the highest point of the shore may be completely covered for only a few days each month, i.e. during spring tides. When the tide is in, the temperature of the surface over which it passes is held more or less constant, there is no dehydration and the oxygen content of the rock pools and other areas of captive water will be upheld. When the tide ebbs the shore is exposed to fluctuating air temperatures and perhaps to the sun's radiation which may quickly heat rocks and sand. Water will be lost by evaporation on hot days with a consequent increase in salinity in rock pools and captive water; conversely when it rains the salinity in pools will fall. Temperature variations will also affect oxygen content. These are just some of the more important problems which face plants and animals living on the shore. Others include exposure and the mechanical effects of waves, which may exert colossal pressures on fixed organisms. There is also the problem of existing on a particular substrate. The type of shore determines the type of flora and fauna to be found. The organisms of an exposed rocky shore will be

▼ The daily rise and fall of the ocean's tides are the result of the gravitational pull of the Moon and that of the Sun, although the effect of the latter is less than half as strong as that of the Moon. The gravitational effect is greatest on the Earth's hemisphere facing the Moon and causes a tidal 'bulge'. When the Sun, Earth and Moon are in line, tide-raising forces are at a maximum, and spring tides occur. At a spring tide, high tide reaches the highest values, and low tide falls to low levels. When lunar and solar forces are not coincidental, with the Sun and Moon at an angle to each other), neap tides occur, which have a small tidal range.

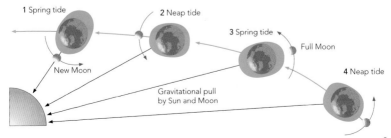

1 Spring tide
2 Neap tide
3 Spring tide
Full Moon
4 Neap tide
New Moon
Gravitational pull by Sun and Moon

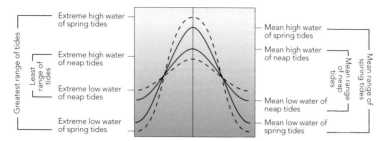

▲ Diagrammatic representation of the tidal range (After Lewis, 1964)

restricted to those that can withstand the effects of waves and wind, but on a sheltered rocky shore we shall find a much greater diversity of species, including those that cannot withstand these mechanical effects.

The manifestation of tidal movements is *zonation* of most plants and animals. Those organisms which can withstand exposure to air, with all its attendant variables, are to be found at the top of the shore where they may be covered by the tide for one or two hours of the day only. Those which can withstand exposure to air for only an hour or so each day will be found at the bottom of the shore. Between these two extremes lies a wealth of different species which can tolerate emersion or immersion to various degrees. Such organisms are zoned according to their requirements, and on a rocky shore or pier side it is usually easy to identify the zones. In each zone a particular type of plant or animal will dominate, and with a little practice they can be recognized easily. Zonation also occurs on sandy and muddy shores, but it is less easily recognized because most of the organisms burrow in the substrate rather than lie on its surface. It should be explained that on an exposed shore the various zones are broader than they are on a sheltered shore. This is because of the way in which wave and wind action serves to extend the influence of sea water up the shore beyond the point that it would have reached if the water were still. The diagram on page 11 shows the scheme of zonation that is recognized in this book.

The theory of zonation was revised by J. R. Lewis. He divided

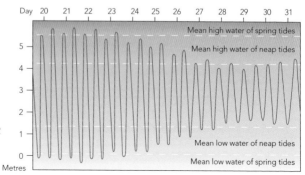

▶ Diagrammatic representation of the spring/neap tide cycle during 12 days at a given point (After Lewis, 1964)

the shore into three biologically defined zones (i.e. defined by the presence of particular species rather than by their height in relation to mean tide level). They are technically named the littoral fringe, the eulittoral zone and the sublittoral zone. These zones are referred to in this book by the following names respectively: *upper shore*, *middle shore* and *lower shore*. Above the upper shore is a region of land that is to some extent under the influence of the sea. Apart from the most sheltered beaches the upper shore is wide, spray extended and above tidal reach. On rocky shores, it is defined as being above the barnacle and terminating at the upper limit of the periwinkles (*Melaraphe neritoides*) and lichens (e.g. *Verrucaria* species). The upper shore is the most difficult zone for marine organisms to live in and usually lacks a great diversity of species. The middle shore is generally a large region, and apart from the most exposed shores it is covered and uncovered by the tides every twelve hours or so. It provides a habitat for the most characteristic shore forms, and there is an abundance of species and individuals. The lower shore is that region extending down from the upper limit of the laminarians and is usually uncovered only at the lowest level of the spring tides. It is the most hospitable zone for marine animals. Newell, R. C. 1970 explains the physiological mechanisms which underlie many of the adaptations shown by intertidal animals. Below the lower shore the shallow sea provides an entirely different set of conditions which are considered on pages 12–14.

## Types of seashore

The European area embraces a great variety of shores depending on the coastal geography and the form of the substrate. Most of these shores can, however, be related to

▼ Scheme of zonation on a rocky shore (After Lewis, 1964)

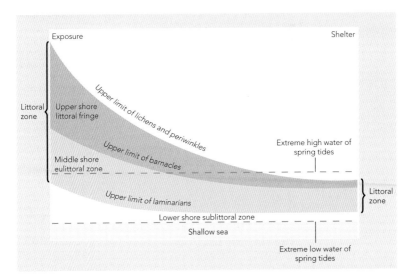

one of four types: rocky, shingle, sandy or muddy. Although one is apt to consider different shore types separately, in practice they often grade into one another. It should be emphasized that at any one point the local climatic conditions and geological formations will have a profoundly modifying effect upon the types of organism to be found.

Rocky shores are the most variable type, their character depending on the prevailing rock. The profile of the rocky shore is usually related to the strata formation. If this is sloping there will usually be a variety of crevices and pools. If it is flat one will encounter rocky platforms and ledges. Rocky shores usually provide a great diversity of animals and plants because the number of ecological niches available is so great.

By comparison with rocky shores all other types of beach may at first sight appear barren and deserted. In practice this is true of shingle beaches where the pebbles are constantly being rolled about by the action of waves, so that most forms of life find it impossible to exist.

Sandy shores are made up of vast numbers of fine grains – usually quartz. The profile of a sandy shore depends to a great extent on the degree of exposure to wave action. Because the particles are so small water is usually retained by capillary action in the minute spaces between the grains. This water effectively lubricates the grains and also allows the animals to survive in the sand after the tide has fallen. The amount of water that is held depends upon the size of the particles, and can be tested by foot pressure. If there is little water in the sand, it appears to look slightly drier around the area of pressure, but if there is much, the sand becomes wet and slushy, especially if you tread up and down several times in one spot. Although the surface of the sand is affected by the fall of the tide, bringing about water loss, temperature and salinity changes, these effects do not appear to penetrate very deeply so that organisms several centimetres down can exist quite well. Providing the sandy shore is stable and not being moved by wave action, it provides a good environment for marine organisms. The diversity of species may not be as great as that on a rocky shore, but the density at which the individuals are disposed on a sandy shore is often very high.

Muddy shores are made of the finest particles of all. For the mud to accumulate the shore must be virtually flat. In addition to the silt (which is of mineral origin) there will also be a variety of organic debris accumulated here. The fine particles may cause difficulty for some animals since it can block delicate structures; some species benefit from the organic material included while others do not. Hence the number of species represented may be restricted to certain specialized ones, although their popula-tions can be considerable. Estuaries with shores of fine sand and mud can be exceedingly productive, as the extent of some commercial cockle beds well demonstrates. Little, C. 2000 provides recent information on sandy and muddy shores.

## The shallow seas

For the purposes of this book the shallow sea is regarded as that area of the sea which stretches from the lowest point to which

**upper shore**   lichen, channelled wrack, small periwinkles, sea slaters

**middle shore**   spiral wrack, serrated wrack, barnacles   *high water*

**lower shore**   kelp, mussel, shore   *low water*
crab, breadcrumb sponge

**Rocky shores** support a wide variety of animal and plant species which are zoned in relation to the tidal height. Green seaweeds are found on the upper shore, brown and red lower down, with the large kelps below average low tide. They provide food and shelter for a range of animals. Rock pools harbour a large variety of animals, some of which cannot survive drying out at low tide.

**upper shore**   sandhopper

**middle shore**   sand mason, rag worm, thin tellin   *high water*

**lower shore** common   *low water*
shrimp, masked crab

**The sandy shore** has a surface which often moves under the action of the tides, so it provides no place for attachment and no inviting crevices, but can hold water between its minute particles. A wide range of animals inhabit the space between the sediment particles at low tide, while others burrow and filter food from the water at high tide.

**upper shore**   green algae

**middle shore**  shore crab   *high water*

**lower shore**   sand gaper, king ragworm,   *low water*
otter shell, cockle

**Muddy shores** have low water movements and are rich in organic matter, often supporting growths of seagrasses that provide a plentiful food supply for grazing animals. As the particles of muddy shores are so fine there is little space between them and oxygen is at low concentration. Burrowing animals, such as clams, may extend siphons above the surface down which a current of water is drawn.

the tide ebbs, to the outer limit of the continental shelf. It will therefore include areas of quite deep water but will not comprise of the vast depths which one associates with the oceans beyond the continental shelf. The extent of the continental shelf is variable, but it takes the form of a gently sloping terrace running seaward. It terminates at the point where the seabed falls away steeply to the ocean depths, and this usually occurs from depths of between 200 and 300m. As the map on page 8 shows, the European land mass is fringed by the continental shelf.

By comparison with the seashore, the shallow sea provides a more constant environment. Organisms living there are not subjected to the changes associated with the ebb and flow of the tides as are those dwelling on the beaches. There is therefore no risk from desiccation, overheating or lack of oxygen. Consequently the shallow sea is populated by a greater diversity of animals including many species not specialized to withstand the rigours of shore life. Plants, on the other hand, are

► *Mactra stultorum* a typical bivalve of shallow water where it burrows in sand and gravel. Its body is enclosed in two shells or valves.

less in evidence because the sunlight upon which they depend for the synthesis of their nutrient materials cannot always penetrate sufficiently through the overlying water. The salinity of shallow seas may be affected by the outfalls of rivers, especially during wet weather, and their mineral content can be influenced by the land mass and water draining from it. Frequently the water is clouded by suspensions of silt swept out to sea near estuaries, and today, sadly, it is often polluted by sewage or chemical waste released from coastal towns and from river mouths.

Whilst diving as a technique has limitations in terms of depth, it alone permits any sort of balanced appreciation of the shallow sea as a habitat. If you cannot dive you cannot get more than a second-hand impression of what is to be found beneath the sea, and you are dependent on vessels equipped with trawls, dredges and underwater television. Essentially two kinds of life are to be found: the floating and swimming organisms which inhabit the waters above the continental shelf; and those animals which live in, or on, the seabed itself. Examples of the former include myriads of drifting, often microscopical organisms forming the plankton, the jellyfishes and sea-gooseberries, a few types of worms and of course the fishes. The animals which live on or in the seabed are mainly invertebrates such as sponges, anemones, molluscs, hosts of worms and a great variety of crustaceans including the familiar crabs and lobsters. In addition, there will be starfishes, sea-urchins and some bottom-dwelling fishes.

▼ *Galathaea intermedia* a squat lobster which lives under stones and rocks on the lower shore and in shallow water. It walks and feeds using jointed appendages.

Just as there are different types of shore, so there are different types of seabed. The texture of the substrate will to a large extent dictate what types of organism live on it, and to a lesser extent the species which live in the water above the substrate, since they may be dependent on the seabed for food, shelter and breeding grounds. Sandy bottoms are characterized by a variety of burrowing invertebrates and the flatfishes. Rocks provide a variety of niches for different species, and shell-gravel and muddy bottoms will also have their characteristic animal life.

## Conservation and collecting

The immediate reaction of many people when they find something of interest is to take it home to have a better look. If possible this should be avoided, and one of the objects of this book is that it should be taken into the field and should help to make removal of specimens less necessary. Although many of the species that are described here are relatively abundant, others are not, and because they may take several years to reach maturity and perhaps have restricted breeding seasons, they can be particularly vulnerable to over collecting. If you must take specimens from the beach try to take as few examples as possible. It is very tempting for divers in particular to collect large numbers of attractive specimens which may at first sight appear abundant; this may not in reality be the case so the temptation should be resisted. If you must take something under water to shoot with take a camera rather than a spear gun. Large, territorial fishes have become severely threatened, particularly in the Mediterranean, due to the use of spear guns, and current opinion is against the use of such devices.

Normally the first sight of any shore gives relatively little impression of the wealth of animal life that may be found there. Most marine plants, on the other hand, are relatively conspicuous. The best time to begin exploring the shore is one hour before low tide is due. When you arrive, go immediately to the water's edge so as to make maximum use of the exposure provided by the ebb tide. You can then work back as the tide returns. On rocky shores you will have to work hard to find representatives of all the animal groups present. Apart from noticing the animals that are attached to the exposed rocks, for instance, barnacles, limpets and dogwhelks, examine all the crevices and holes for periwinkles, topshells and sea-anemones. Where there is a good covering of seaweeds notice their zonation patterns and turn them over to look for the variety of species that may be growing on or under them. You should find hydroids, sponges, sea-mats and periwinkles at least. If there are smaller rocks which are free, turn them over and look for sea-anemones, worms, crabs and other crustaceans below. The animals you have exposed have deliberately chosen the underside of rocks, so ensure that you *turn the rocks back again after examination,* otherwise they will probably die. Investigate all rock pools carefully. They frequently contain a variety of organisms including swimming forms such as prawns and small fishes. On sandy shores and on mud, signs of life may at first sight be less obvious. The experienced observer will soon recognize the various types of worm cast and burrow entrance, however. To find what lives within the sand or mud you will have to dig carefully and sometimes sieve the sand. Such shores house a wealth of burrowing worms and bivalves as well as a few burrowing sea-anemones, snails, crustaceans, brittle-stars and heart-urchins.

If you are collecting for a purpose, arrange to look after your specimens carefully. Prevent them from becoming overheated by placing your jars in a rock pool to keep the temperature down. Put the lids on only when actually carrying the jars, and add just sufficient water to cover the specimens or let them swim freely. Only take the animals away from the beach if you are certain that you can keep them alive. After you have examined them return them to the spot they came from, or to a similar place. Much of what has been said applies to the diver collecting in the shallow seas. If fishes are being collected from 10m or below, remember that they will have to be decompressed by gradually raising them to the surface. Polythene bags are useful for this and they also make good containers for small invertebrates; larger specimens can be carried in string bags.

# ILLUSTRATED KEY

## Plants

Many minute filaments

*24–27, 30–31,*
*36–37, 40–41,*
*52–55, 58–61*

Leaf-like

*28–29, 34–35,*
*56–57*

Hard, chalky and encrusting

*50*

Coarse filaments

*22–23, 28–29,*
*34–35, 42–45,*
*50–51, 58–59*

Frond without midrib, but with side-branches

*42–47,*
*50–53, 56–57*

Branching felt-like rounded fronds

*24–25*

Collapsed green tube

*22–23, 25*

Frond with midrib

*34*

Delicate and fan-like

*22–23, 37,*
*48, 52–53, 56–57, 61*

Branching and ribbon-like, without midrib

*30,*
*36–37,*
*42–43, 46–51, 56–57*

Large, with wide frond

*32–33*

Flat and encrusting, smooth or rough

*26, 48, 50, 62–63*

Branching and ribbon-like, with midrib

*36, 38–39, 59*

Spherical or club-like

*27–29*

Strap-like leaves

*62, 64–65*

16

# Plant-like animals

Simple, branching or encrusting spongy growths

*70–75*

Branching growths; tentacles cannot withdraw

*78–79*

Branching growths; tentacles can withdraw

*81–82*

Soft body; tentacles surround mouth

*88–97, 100*

Hard body; soft tentacles surround mouth

*100*

Hard, branching growth: soft tentacles

*101–102*

Fan-like growth

*88–89, 96–97, 102*

Free-living; 5 pairs of branching arms

*252*

Branching colony: individual zooids minute and only visible under a lens. Tentacles withdraw if disturbed

*244–247, 214–215*

Encrusting colony; individual zooids minute and only visible under a lens. Tentacles withdraw if disturbed

*244–249*

# Joint-legged animals

Small body; 6–8 legs

*240–241*

Flat or compressed body; more than 8 legs

*206–220*

Rounded or compressed body; with pincers; more than 8 legs

*220–224*

Moderate–large, long body; with pincers; walking legs

*206, 225–231*

Small–large, ovoid or rounded body; with pincers; walking legs

*228, 232–239*

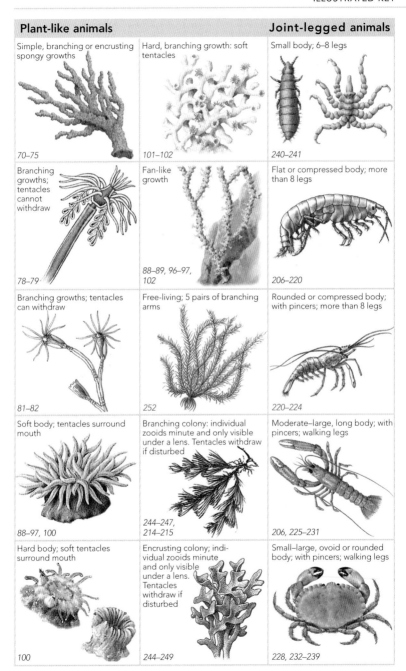

17

## Animals with shells

| | | |
|---|---|---|
| With 8 shell plates | Shell-like tube | 2 small, hinged shells in chalky tube |
| 144–145 | 136–139, 156 | 195 |
| Flat, perforated shell | Cap or slipper-like shell | Star-like body |
| 146 | 157 | 252–259 |
| Cone-like shell | Shell like tiny elephant's tusk | Rounded, spiny body |
| 147–148 | 173 | 260–265 |
| Spiral, ungrooved shell | 2 Equal-shaped shells hinged together | Stalked, flat body, with shell-like plates |
| 149–155, 162–163, 172–173 | 174–195 | 202 |
| Spiral, grooved shell | 2 Unequal-shaped shells hinged together | Small, volcano-like shell; attached |
| 156, 160–161 | 176–179, 182 | 203–205 |

## Animals without shells

Small leaf-like worm

*106–107*

Worm with head or tail projections

*140–141*

Body vase or flagon-like; attached sometimes in colonies

*272–275*

Long, unsegmented worm

*108–111, 269*

Body jelly-like

*84–87, 104–105, 270–271*

Body shark-like

*278–281*

Segmented worm without conspicuous gills or body tentacles

*116–131*

Body slug-like

*164–171*

Body ray-like

*283*

Segmented worm with gills or tentacles

*118–119, 124–125, 128–135*

Body soft; 8–10 tentacles with suckers, eyes conspicuous

*196–199*

Body eel-like

*286–287*

Tubed worm; with head fan

*135–138*

Body cucumber-like

*266–267*

Bony fish

*294–313*

The plant kingdom is divided into a number of major groups which range from simple, unicellular organisms that can only be seen with the aid of the microscope, to the great trees which are so familiar on land. The sea provides a satisfactory environment for only a few groups of plants, however, and the majority of these belong to the lowest group of the plant kingdom, the Thallophyta. This group comprises the algae (including the seaweeds), the fungi and the lichens. The only other group that is significantly represented in the sea is the Angiospermae (flowering plants).

Green plants are anabolic organisms, that is, they elaborate organic material from inorganic sources with the aid of photosynthesis. No animal can do this; all animals depend either directly or indirectly on plants for their supply of food. Plants therefore form the first living stage of all *food chains* or *food webs*. (Food chains or food webs are terms used to describe the nutritional relationships existing between organisms and their consumers or predators in a community.)

The marine algae are divided into a number of subgroups of which three will be dealt with here. They are the Chlorophyceae (green algae), Phaeophyceae (brown algae) and Rhodophyceae (red algae). Whilst colour would appear to be their principal distinguishing characteristic, it must be remembered that colour may not only be variable, but it may change considerably if the plants have been removed from the sea for long periods, or if they are dying. Furthermore, many of the red and brown seaweeds are quite similar in colour. These facts make colour alone a rather unreliable criterion for the purposes of identification, and this also holds true for animals. The identification of marine algae is largely based on the form of the plant in question. Precise identification may depend on the accurate recognition of microscopic characters, and to a large extent such aspects of seaweed identification are beyond the scope of this book. Nevertheless with the gross descriptions of form, the colour plates and a little experience, it should be possible for the reader to succeed in what is quite a difficult area of seashore biology.

▼ *Palmaria* shows the typical seaweed form of holdfast and frond.

A typical seaweed is normally composed of a *holdfast* (or attachment organ) and a *frond*. The holdfast and frond together are termed the *thallus*. Many fronds become broken from their holdfasts and washed into positions where they are subsequently found. For this reason not all the illustrations show the holdfast. Unlike the higher plants, many algae show relatively little differentiation within their tissues, and only some, such as the laminarians, have any form of vascular system. Instead, most of the cells are developed in such a way that each can carry out many of the functions of the organism, and any specialized regions that do exist, such as holdfasts and reproductive organs, are reduced to a minimum. The form of the frond and the holdfast are important clues to identification. Holdfasts may be root-like, disc-like or may grow as a shaggy tangle of fine 'rootlets' – these are not roots in the strict sense, however, as they do not provide the seaweed with a special absorptive region. The frond may

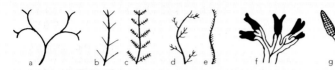

be rounded or cylindrical, flattened, flattened with a central rib or ribs, or intermediate, i.e. neither round nor flat. The style of branching is also important; it may be dichotomous, alternate, spiral or pinnate. The diagram above shows some of these features.

In some species, especially the brown and red algae, attention to the form and disposition of the reproductive organs will assist with identification. In the Phaeophyceae this is true of the order Fucales (e.g. *Fucus*). Here, the reproductive organs generally consist of small pits opening by minute pores. They are usually set in thickened portions of the frond near the tips of the branches. In other orders, such as the Ectocarpales (e.g. *Ectocarpus*), reproductive organs can be identified (with a hand lens) on the sides of the branches or in the angle of a junction between the stem and a side-branch. In some other genera (e.g. *Padina*), they occur on the surface of the frond arranged in groups. Some of the Rhodophyceae also have their reproductive organs set in pits. Other genera, like *Polysiphonia*, carry them on small branches of the frond, but they may be buried in the plant tissue itself, as in *Dilsea*. In some red algae the reproductive organs are extremely difficult to discern. The diagram above also illustrates some reproductive organs.

The occurrence of seaweeds is limited by light, which they require for photosynthesis. As has already been mentioned, zonation of organisms is a fundamental characteristic of the shore. Generally the green algae grow high on the rocky shore, either where they can be exposed as the tide ebbs, or in pools. The brown algae frequently form a characteristic order down the shore, i.e. *Pelvetia canaliculata*, *Fucus spiralis*, *Ascophyllum nodosum*, *F. vesiculosus*, *F. serratus* and *Laminaria* species. The red algae generally prefer less exposure to air and grow to the greatest depths in the sea. It must be remembered that the lists of descriptions of algae which follow are by no means exclusive. It has not been possible to include many of the smaller genera existing in the European area, and many that are mentioned are in fact represented by more species than can be described here. A very helpful check-list of British marine algae is given by Parke, M. and Dixon, P. S. 1976.

This account has paid little attention to the lichens or the marine angiosperms. However, more is said about them on pages 62–65. Here it should be noted that the former group dwell in the splash zone and are more characteristic of the terrestrial habitats. The sea-grasses are highly specialized angiosperms.

▲ Various patterns of thallus branching and reproductive organs: **a** dichotomous; **b** opposite; **c** opposite with pinnate side-branches; **d** alternate with alternate side-branches; **e** spiral branching; **f** sprig of **Fucus** with reproductive area shaded in black; **g** magnified reproductive bodies of **Padina**; **h** magnified reproductive body of **Polysiphonia**

▼ Some seaweeds like *Ectocarpus* comprise many fine filaments grouped together.

These mainly aquatic plants contain the green pigment chlorophyll. Their bodies are not differentiated into true roots, stems nor leaves, and they lack a vascular system. Reproduction is by spores.

## Class Chlorophyceae Green algae

Algae in which the chlorophyll is not masked by brown or red pigment. Their structure ranges from minute, single-celled plants to larger thread- or frond-like plants composed of many cells.

Prasiola
stipitata

### Prasiola stipitata Suhr
**Frond** 7–12.5mm long; generally ovoid narrowing to a short stalk; edges often curled.
**Colour** usually dark green.
**Habitat** on rocks on the upper middle shore; often associated with bird guano.
**Distribution** Atlantic, English Channel and North Sea. N.B. several other species occur.

Monostroma
grevillei

### Monostroma grevillei (Thuret) Wittrock
**Frond** 100–300mm long; soft, delicate and funnel shaped with a split down one side when adult.
**Colour** pale, translucent green.
**Habitat** usually in rock pools on the lower shore or on rocks in shallow water.
**Distribution** Atlantic, English Channel, North Sea and Baltic. N.B. about five other species also occur.

### Blidingia minima (Kützing) Kylin ( =Enteromorpha minima)
**Frond** up to 100mm long: maybe unbranched or branched slightly; sometimes inflated.
**Habitat** growing on rocks, pilings, etc., on the upper shore.
**Distribution** Atlantic, English Channel and North Sea. N.B. several other species occur. Distinguishing between the species of Enteromorpha is a specialist's task. Identification to genus should suffice in most cases.

Blidingia
minima

### Enteromorpha intestinalis (Linnaeus) Link
**Frond** 50mm–1m or more long; tubular and irregularly inflated and crinkled; generally not branched; may taper continuously along length.
**Colour** pale green.

Enteromorpha
intestinalis

**Habitat** in rock pools on upper shore, often in water with reduced salinity; sometimes washed up.
**Distribution** Mediterranean, Atlantic, English Channel, North Sea and Baltic.

### *Enteromorpha compressa* (Linnaeus) Greville
**Frond** up to 300mm long; tubular; main frond tapers only at the base and usually gives off a number of similar side-branches; not as soft and delicate as *B. minima* above.
**Colour** dark to pale green.
**Habitat** on rocks and pilings on upper, middle and lower shores; frequently in pools where the salinity is reduced; often washed up; sometimes in estuaries.
**Distribution** Mediterranean, Atlantic, English Channel, North Sea and Baltic. N.B. this may be a branched form of *E. intestinalis*.

### *Enteromorpha linza* (Linnaeus) J. Agardh **( =Ulva linza)**
**Frond** 100–500mm long; superficially flattened, but will be found to be hollow if a transverse section is carefully examined microscopically at the edges; unbranched, tapering slightly from middle towards apex and base; edges crinkled.
**Colour** bright green.
**Habitat** on rocks on upper, middle and lower shore; sometimes washed up.
**Distribution** Mediterranean, Atlantic, English Channel, North Sea and Baltic. N.B. about eight other species also occur.

### *Ulva lactuca* Linnaeus
**Sea-lettuce**
**Frond** from 150–500mm long; variably shaped, may be lobe-shaped, lance-shaped or perforated; generally wider at the top than at the base; stalk if present is solid.
**Colour** translucent green.
**Habitat** on rocks on upper, middle and lower shore; sometimes in pools; sometimes in shallow water; may float free or be washed up.
**Distribution** Mediterranean, Atlantic, English Channel, North Sea and Baltic. N.B. about three other species also occur.

*Enteromorpha
linza*

*Enteromorpha
compressa*

*Ulva
lactuca*

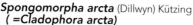

### Spongomorpha arcta (Dillwyn) Kützing ( =Cladophora arcta)

Frond about 70mm long; forms hemispherical tufted growths of many associated branching threads which are held together by minute coils and hooks; some branches grow down to assist with attachment.

Colour usually deep green.

Habitat on rocks and stones, middle and lower shore.

Distribution Atlantic, English Channel, North Sea and Baltic. N.B. common in Spring and early Summer. Several other species occur.

*Spongomorpha aeruginosa*

### Spongomorpha aeruginosa (Linnaeus) Hoek ( =Cladophora lanosa)

Frond 10–30mm long; forming growths of many woolly looking threads to give a spherical, tuft-like appearance.

Colour light to dark green.

Habitat attached to other seaweeds and the sea-grasses (e.g. *Zostera,* see pages 64–65).

Distribution Atlantic, English Channel, North Sea and Baltic.

*Spongomorpha arcta*

### Chaetomorpha linum (O. F. Müller) Kutzing

Frond 150–300mm long forming growths of many fine, almost cylindrical, unbranching threads which taper slightly at their bases (a hand lens will show that each filament is made of a chain of large cells), threads fixed by modified cell at base which often merges with that of neighbouring threads to form an attachment disc.

Colour mid green.

Habitat in pools on upper and middle shore, or attached to rocks covered by sand; often floating free and washed up.

Distribution Mediterranean, Atlantic, English Channel, North Sea and Baltic. N.B. several other species occur.

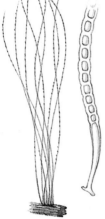

### Cladophora rupestris (Linnaeus) Kützing

Frond 70–120mm long; forming small to large coarse growths of many branching fronds; branching is usually irregular or opposite (see page 21); attachment area may send off runners to start new growths and so colonize large areas.

Habitat on rocks on the middle and lower shore where it may grow underneath *Fucus serratus* (see page 38).

Distribution Atlantic, English Channel, North Sea and Baltic. N.B. a closely related species occurs in the Mediterranean. About fifteen other species occur.

*Chaetomorpha linum*

**Chaetomorpha melagonium** (Web and Mohr) Kützing
(not illustrated) Similar to *Chaetomorpha linum* (opposite).
Frond up to 300mm long.
Colour dark green.
Habitat lower shore and shallow water.
Distribution widespread.

**Derbesia marina** (Lyngbye) Solier
Frond about 50mm long; usually forms a
growth of fine filaments rising from
a basal portion with occasional
lateral branches; sometimes with
ovoid reproductive bodies that
can be seen with a hand lens.
Colour bright green.
Habitat on seaweed and mud on the
upper, middle and lower shores and in
shallow water.
Distribution Atlantic and English Channel.

*Derbesia
marina*

*Cladophora
rupestris*

**Bryopsis plumosa** (Hudson)
C. A. Agardh
Frond up to 100mm long; feather-like,
pinnate branches arranged more or less
opposite on the main stem; branches
are usually smaller towards the top.
Colour yellow green in the male,
and dark green in the female,
glossy.
Habitat on rocks and stones
and on pool sides, middle
and lower shore and
in shallow water.
Distribution
Mediterranean,
Atlantic, English
Channel, North
Sea and Baltic. N.B.
common in Spring
and Summer. One
related species occurs.

*Bryopsis
plumosa*

**Codium tomentosum** Stackhouse
Frond 250–350mm long; tubular and dichotomously
branching with a felt-like texture.
Holdfast disc-like, consists of many entwined threads
encrusting the substrate.
Colour dark green.
Habitat on mud, sand and rocks down to 20m.
Distribution Mediterranean, Atlantic and English Channel.
N.B. several other species occur.

*Codium
tomentosum*

# Class Phaeophyceae Brown algae

*Ectocarpus*
species aggregate

Algae in which the chlorophyll is often masked by the brown pigment fucoxanthin. These are multicellular plants, often large, and are normally attached to the substrate. They do not generally flourish in warmer water, thus they are relatively scarce in the Mediterranean. For more details see Hiscock, S. 1979.

### *Ectocarpus* species aggregate
**Frond** 120–300mm long, tangled growth of fine, branching filaments which becomes free towards the tips; branches very variably arranged.

**Holdfast** creeping and filamentous, a hand lens may reveal both club-shaped and more pointed reproductive bodies which are carried towards the tips of the branches, usually on short stems.

**Colour** yellow-green-brown.

**Habitat** attached to rocks and stones from the middle shore down to shallow water.

**Distribution** Mediterranean, Atlantic, English Channel, North Sea and Baltic. N.B. this represents an aggregation of ten or so similar and related species, extremely difficult to separate in the field.

*Spongonema*
*tomentosum*

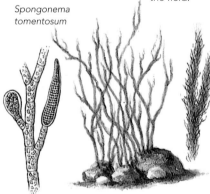

### *Spongonema tomentosum*
(Hudson) Kützing
### ( =*Ectocarpus tomentosum*)
**Frond** 30–130mm long, but often less; similar to *E. siliculosus* above, but plants more entwined thus appearing as corded, woolly growths about 2.5mm wide.

**Habitat** growing on larger seaweeds, e.g. *Fucus vesiculosus* or *Himanthalia* sp. (see pages 35 and 39), or on stones and shells on middle and lower shores.

**Distribution** Atlantic, English Channel and North Sea. N.B. five or so closely related species very difficult to distinguish in the field.

detail of a filament
with reproductive
bodies

### *Ralfsia* species aggregate (Areschoug) J. Agardh
Groups of individuals forming irregular encrustations 20–100mm across; individuals may be round and about 2.5mm thick; in the winter small, club-like reproductive bodies growing upwards may be seen with a hand lens; isomorphic generations (see page 21) follow each other in the life cycle.

*Ralfsia* species
aggregate

**Colour** dark brown-black.

Habitat attached to rocks and shells, often in quite exposed places.
Distribution Mediterranean, Atlantic, English Channel and North Sea. N.B. about four other species occur.

### Leathesia difformis (Linnaeus)
**Areschoug**
Frond 20–50mm in diameter; globular or irregularly shaped, shiny, thick-walled growths which are solid when young but become more hollow with age; if the growths are cut across with a sharp knife and the inside examined with a hand lens it will be seen to be made of dichotomously branching filaments; the reproductive bodies are carried terminally on the outer extremities of these.
Colour olive-brown.
Habitat growing on rocks and smaller seaweeds.
Distribution Atlantic, English Channel and North Sea in Spring and Summer. N.B do not confuse with *Colpomenia* (see page 29).

*Leathesia difformis growing on another alga*

### Stilophora rhizodes
(Turner) J. G. Agardh
Frond 150–600mm long; dichotomously branching frond (see page 21); branches taper near their tips and are covered in small spots; frond is solid when young but becomes tubular with age.
Habitat on rocks or other seaweeds on the lower shore and shallow water, often where the salinity is reduced, e.g. by streams.
Distribution Atlantic and English Channel.

*Stilophora rhizodes*

### Stictyosiphon tortilis (Ruprecht) Reinke
Frond 70–150mm long; much branched, alternately or opposite (see page 21).
Colour yellow-brown.
Habitat on stones, other seaweeds and shells on the lower shore and in shallow water.
Distribution north Atlantic and northern North Sea. N.B. several other species occur.

*Strictyosiphon tortilis*

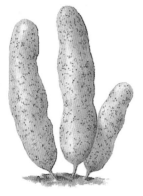

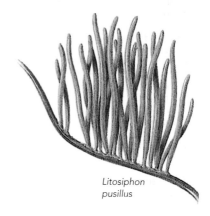

Litosiphon
pusillus

Asperococcus
turneri

### Asperococcus turneri (Smith) Hooker ( =A. bullosus)
Frond 150–300mm long; tubular growth carried on a narrow,
short stalk.
Holdfast small and disc-like. Often growing in groups; texture
soft, slightly transparent and membranous, thickening with age.
Colour olive green.
Habitat attached to rocks and larger seaweeds on the middle
and lower shore and in shallow water.
Distribution Mediterranean, Atlantic, English Channel and
North Sea in Summer. N.B. several other species occur.

### Litosiphon pusillus (Hooker) Harvey
Frond 50–100mm long; densely growing in clumps; texture
soft and elastic; young specimens may be slightly hairy, older
ones tubular.
Habitat growing on Chorda filum (see page 34),
which may thus resemble frayed ropes at the end of
the growing season, on the extreme lower shore
and in shallow water.
Distribution Atlantic, English Channel and North
Sea. N.B. several other species occur.

### Scytosiphon lomentaria (Lyngbye) Link
Frond 150–300mm long; tubular and resembling a
miniature chain of sausages; not branched; tapers at
the tip.
Colour green-yellow.
Habitat attached to rocks, stones and other seaweeds
on the lower shore and in shallow water, often in
exposed places.
Distribution Mediterranean, Atlantic, English Channel,
North Sea and Baltic.

Scytosiphon
lomentaria

### *Punctaria* species aggregate Greville

**Frond** 200–400mm long and 75mm across unbranched, leaf-like fronds borne on short stalk and dotted with small spots or hairs: may terminate in a wide-or narrow-angled point.

**Habitat** on rocks, stones and shells (especially those of live limpets see pages 147–148) on middle and lower shores and in shallow water.

**Distribution** Mediterranean, Atlantic, English Channel and North Sea. N.B. about three other species occur.

### *Petalonia* species aggregate (O. F. Müller) Kuntze

**Frond** may reach 300mm in length and 60mm across; edges may be frilled; the short stalk quickly widens to form the frond and is much shorter than that of the laminarians (see pages 32–33).

**Colour** glossy yellow-brown.

**Habitat** on rocks, often those covered with sand, and in pools on the lower shore.

**Distribution** Mediterranean, Atlantic, North Sea and Baltic. N.B. about two other species occur.

### *Colpomenia peregrina* Sauvageau
**Oyster Thief**

**Frond** is a thin-walled, globular, hollow growth up to 200mm in diameter, although often the size of a table-tennis ball; covered all over with fine brown dots.

**Habitat** in pools and attached to various seaweeds and shells on middle and lower shores and in shallow water.

**Distribution** Atlantic, English Channel and North Sea. N.B. this species is easily distinguished from *Leathesia difformis* (see page 27), since it is not gelatinous but is dry and papery.

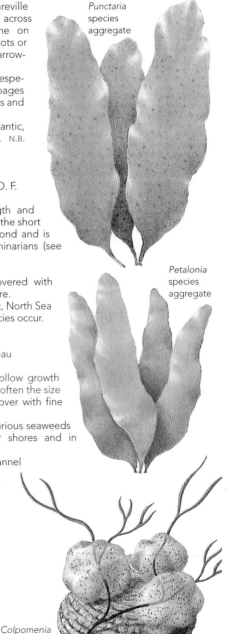

*Punctaria* species aggregate

*Petalonia* species aggregate

*Colpomenia peregrina*

29

### Cutleria multifida (Smith) Greville

**Frond** 100–400mm long; forms flat, fan-like growths; dichotomously branched, tips of the branches divided; texture rather springy when fresh; usually spotted.

**Holdfast** disc-like.

**Colour** yellow-green.

**Habitat** attached to rocks and shells, generally in shallow water, may be washed up. The thick, encrusting aglaozonia phase (sporophyte) of *Cutleria multifida* is common in shallow water in the north-eastern Atlantic.

**Distribution** Mediterranean, Atlantic, English Channel and North Sea.

Aglaozonia phase

*Cutleria multifida*

### Sporochnus pedunculatus

(Hudson) C. A. Agardh

**Frond** 150–450mm long; filamentous growth of a central thread, bearing branches; each branch bears alternately arranged branchlets.

**Colour** olive-green.

**Habitat** on rocks on lowershore and in shallow water.

**Distribution** Mediterranean, Atlantic, English Channel and North Sea.

*Sporochnus pedunculatus*

### Arthrocladia villosa

(Hudson) Duby

**Frond** 150–900mm long; consists of fine filaments with oppositely arranged branches; these in turn bear branchlets set in whorls (as shown by inset); branchlets may carry very fine, unbranching filaments which form part of the reproductive bodies and these impart a shaggy, green appearance.

**Habitat** growing on rocks, stones and *Zostera* (see pages 64–65).

**Distribution** Mediterranean, Atlantic, English Channel and North Sea. N.B. distinguished from *Desmarestia* (see opposite) by the shaggy appearance resulting from the whorled branchlets.

*Arthrocladia villosa*

### *Desmarestia aculeata* (Linnaeus) Lamouroux
**Frond** 300mm–1.8m long; main stem is slightly compressed and pliable when young, becoming more rigid with age; main stem bears alternately arranged side-branches which have a thorny appearance; side-branches themselves are alternately pinnate (see page 21) and in Summer they bear very fine, branching filaments.
**Holdfast** disc-like.
**Habitat** attached to rocks on extreme lower shore and in shallow water; sometimes washed up.
**Distribution** northern Atlantic and northern North Sea N.B. other species of *Desmarestia* rot quickly when taken from the sea; *D. aculeata* only does so in Summer. Their colour changes from bright green (when young) or brown (when older) to verdigris.

*Desmarestia aculeata*

### *Desmarestia ligulata* (Lightfoot) Lamouroux
**Frond** 500mm–1.8m long; main stem somewhat flattened with a suggestion of a midrib, and bearing closely arranged, opposite branches; these branches themselves bear minute projections which may carry tufts of fine hairs.
**Colour** olive-brown.
**Habitat** attached to rocks in pools and in shallow water.
**Distribution** Atlantic and English Channel. N.B. the pliable stem decomposes in the air to a flabby, green condition.

*Demsarestia ligulata*

### *Desmarestia viridis* (O.F. Müller) Lamouroux
**Frond** about 300mm long; shorter and more delicate than *D. aculeata* (above); soft, pliable, slightly flattened stem carries oppositely arranged side-branches which are longer near the base and shorter away from it. These side-branches carry opposite sub-branches which give the plant a feathery appearance, and this opposite branching (see page 21) distinguishes this species from *D. aculeata* where the branching is alternate.
**Colour** olive-green when young, red-brown when older.
**Habitat** attached to stones and seaweed on the lower shore and in shallow water.
**Distribution** Atlantic, English Channel and North Sea in Summer.

*Desmarestia viridis*

31

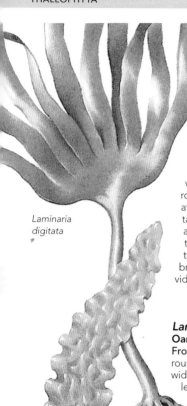

Laminaria
digitata

Laminaria
saccharina

Note on the family Laminariaceae The seaweeds illustrated on this plate are all members of an important family of brown algae, the Laminariaceae (kelps). This family includes some of the largest algae found in European waters. There is a characteristic alternation of heteromorphic generations (see page 21) and it is the asexual plant (sporophyte) which forms the typical kelp plant. The sexual phase (gametophyte) is microscopic and generally unfamiliar to the layman. Laminarians are characteristic of the extreme lower shore and of shallow, coastal water. They form a very characteristic zone on many rocky shores where they are usually uncovered only at the Spring tides. Laminarians provide an important habitat for a variety of other marine algae and animals. Some of these are specialized to live on the frond, but many more find food and shelter in the crevices and crannies provided by the massed branches of the holdfast. These holdfasts thus provide a rich hunting ground for the marine zoologist.

### *Laminaria digitata* (Hudson) Lamouroux
**Oarweed** or **Tangle**
**Frond** usually about 1m, but may be longer; thick. rounded, very flexible stipe broadening out into a wide, shiny blade; blade and stalk are about equal in length; stipe floppy and oval in cross section; the blade divides as it grows forming a number of strap-like fronds. Plants growing in exposed places generally have more fronds than those in sheltered ones.
**Holdfast** much branched.
**Habitat** attached to rocks on the extreme lower shore and down to about 1m; often washed up.
**Distribution** Atlantic, English Channel, North Sea and Baltic. N.B. the brown weed fades to green and then white when washed up on the beach; this species usually occupies a slightly higher position on the extreme lower shore than does *L. hyperborea* (see below).

### *Laminaria hyperborea* (Gunnerus) Foslie
**Frond** up to 3.5m long but often less; stiff, rough, rounded stalk tapers towards its upper part; blade is ovoid and split into a number of strap-like fronds. Stipe stiff and round in cross section.
**Holdfast** much branched.
**Habitat** attached to rocks and stones on the extreme lower shore and in shallow water.

**Distribution** Atlantic, English Channel, North Sea and Baltic.
N.B. the roughness of the stalk provides good anchorage for
epiphytic and epizoic organisms; usually growing just below
*L. digitata* (see opposite).

*Laminaria
hyperborea*

### *Laminaria ochroleuca* De la Pylaie
(Not illustrated) Similar to *L. hyperborea* (opposite), but
blade yellowish and stem stiff and smooth.
**Distribution** Atlantic north to English Channel
approaches.

### *Laminaria saccharina* (Linnaeus) Lamouroux
**Sea Belt, Sugar Kelp** or **Poor Man's Weather Glass**
**Frond** 200mm–3m long; relatively thin stalk which is
usually about one quarter the blade length; blade
somewhat like a crumpled ribbon.
**Holdfast** of branching growths and two-tiered in
appearance.
**Habitat** attached to stones, rocks and shells from
extreme lower shore down to about 20m, especially
in sheltered positions.
**Distribution** Atlantic, English Channel and North
Sea. N.B. avoid confusion with *Petalonia* species
aggregate (see page 29).

### *Saccorhiza polyschides* (Lightfoot)
Batters **( =S. bulbosa)**
**Frond** 1.5–4.5m long and nearly as broad
across the blade; stalk or stipe is flat and
twisted at the base, being wavy at the
edges and quite stiff towards the upper
part; blade widens suddenly from top of
stipe and is almost semicircular; minute
hairs borne in tufts on blade.
**Holdfast** is collar-like or bulbous, with a
number attachment 'rootlets'.
**Habitat** on rocks on extreme lower
shore and in shall water.
**Distribution** Atlantic, English Channel and North Sea.

*Saccorhiza
polyschides*

33

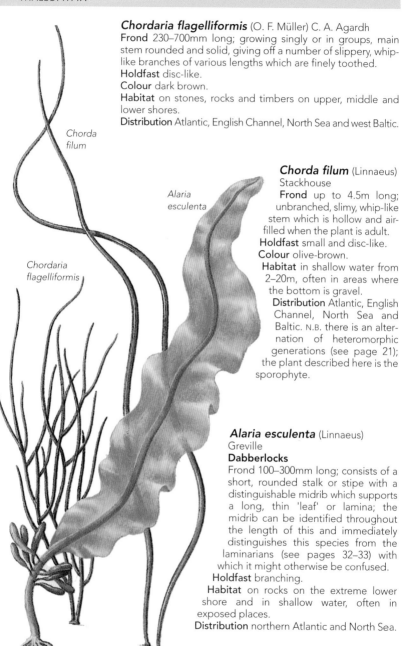

**Chordaria flagelliformis** (O. F. Müller) C. A. Agardh
Frond 230–700mm long; growing singly or in groups, main stem rounded and solid, giving off a number of slippery, whip-like branches of various lengths which are finely toothed.
Holdfast disc-like.
Colour dark brown.
Habitat on stones, rocks and timbers on upper, middle and lower shores.
Distribution Atlantic, English Channel, North Sea and west Baltic.

*Chorda filum*

*Alaria esculenta*

*Chordaria flagelliformis*

**Chorda filum** (Linnaeus) Stackhouse
Frond up to 4.5m long; unbranched, slimy, whip-like stem which is hollow and air-filled when the plant is adult.
Holdfast small and disc-like.
Colour olive-brown.
Habitat in shallow water from 2–20m, often in areas where the bottom is gravel.
Distribution Atlantic, English Channel, North Sea and Baltic. N.B. there is an alternation of heteromorphic generations (see page 21); the plant described here is the sporophyte.

**Alaria esculenta** (Linnaeus) Greville
**Dabberlocks**
Frond 100–300mm long; consists of a short, rounded stalk or stipe with a distinguishable midrib which supports a long, thin 'leaf' or lamina; the midrib can be identified throughout the length of this and immediately distinguishes this species from the laminarians (see pages 32–33) with which it might otherwise be confused.
Holdfast branching.
Habitat on rocks on the extreme lower shore and in shallow water, often in exposed places.
Distribution northern Atlantic and North Sea.

### *Mesogloia vermiculata* (Smith) S. F. Gray
**Frond** about 60mm long; slimy, rounded stem bears numerous branches of various thicknesses and lengths giving them a tufted, fan-like, pyramidal appearance.
**Colour** green-yellow-brown.
**Habitat** on rocks often covered by sand on the middle and lower shores.
**Distribution** northern Atlantic and North Sea. N.B. several other species occur.

*Mesogloia vermiculata*

### *Himanthalia elongata* (Linnaeus) S. F. Gray
### ( =H. lorea)
**Seathong**
**Frond** up to 2m long; strap-like, flattened and dichotomously branched; branches taper towards their tips; the 'main' portion of the plant grows up from a button-like structure; the buttons, which may often exist without the rest of the frond, develop just above the holdfast and are a useful distinguishing feature.
**Colour** green-brown.
**Habitat** growing in colonies in pools on the lower shore or in shallow water.
**Distribution** Atlantic, and English Channel.

### *Sargassum muticum* (Yiendo)
**Japweed or Wireweed**
**Frond** up to 4m long, with regular alternate side branches which hang down from the main stem if this is held horizontally, like washing from a line. Each branch has flattened oval 'leaflets' and round bladders.
**Colour** brown.
**Habitat** in shallow water attached to rocks and boulders.
**Distribution** Irish Sea, English Channel and southern North Sea. An invasive species introduced from Japan.

*Himanthalia elongata*

*Sargassum muticum*

*Halopteris scoparia*

### *Halopteris scoparia* (Linnaeus) Sauvageau ( =*Stypocaulon scoparium*)

**Frond** 80–150mm long; shaggy appearance in Summer, but less so in Winter; main stem branches alternately so that the tufted branches resemble a series of cones arranged upside-down on top of one another.
**Colour** dark brown.
**Habitat** attached to rocks on the lower shore, in rock pools and in shallow water.
**Distribution** Mediterranean, Atlantic and English Channel.

### *Cladostephus verticillatus*
(Lightfoot) C. A. Agardh

**Frond** 100–250mm long; main stem more or less dichotomously branching with branches bearing whorled, spiny branchlets; these may be lacking in the lower regions of the main stem and principal branches.
**Holdfast** disc-like.
**Colour** generally dull brown.
**Habitat** on rocks, stones and red coralline seaweeds on the middle and lower shores.
**Distribution** Mediterranean, Atlantic, English Channel and North Sea.

*Cladostephus verticillatus*

*Halopteris filicina*

### *Halopteris filicina*
(Grateloup) Kützing

**Frond** 50–100mm long; main stem bears many alternately arranged side-branches which themselves carry pinnate branchlets (see page 21); usually the upper part of the stem bears more branches than the lower.
**Holdfast** root-like.
**Colour** green-brown.
**Habitat** on rocks, larger seaweeds and shells on the lower shore.
**Distribution** Mediterranean, Atlantic and English Channel.

### *Dictyopteris membranacea*
(Stackhouse) Batters

**Frond** 100–300mm long; dichotomously branching, flattened; thinner than those of *Fucus* (see pages 38–40); a conspicuous midrib may be the only part of the stem left in the lower regions of older specimens; membranous edges of fronds are dotted with groups of minute hairs; tips of branches are rounded and slightly split or notched.
**Holdfast** disc-like and fibrous.

*Dictyopteris membranacea*

Colour yellowish when juvenile, growing to darker brown.
Habitat on rocks on the extreme lower shore and down to 80m.
Distribution Mediterranean, Atlantic north to south-west
Britain and Ireland, and western English Channel. N.B. when
freshly collected it has a most unpleasant odour.

### *Dictyota dichotoma* (Hudson) Lamouroux

Frond about 130mm long; regularly dichotomous;
delicate and flattened with rounded, bifid tips; no
midrib; fronds may be covered with groups of
minute, hair-like reproductive bodies.
Colour yellow-olive-brown.
Habitat on rocks and other seaweeds on
middle and lower shores.
Distribution Mediterranean, Atlantic,
English Channel and North Sea.

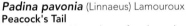

*Dictyota
dichotoma*

### *Padina pavonia* (Linnaeus) Lamouroux
**Peacock's Tail**

Frond about 100mm long; fan-shaped, narrow,
rounded stalk or stipe gives rise to a rounded lamina;
when young and smaller this is often quite thin and flat, but as it
matures it develops into the characteristic concave fan.
Colour outer surface has brown-green stripes; inner
surface is lime-green.
Habitat on stones and rocks in shallow water.
Distribution Mediterranean, Atlantic and
English Channel.

*Padina
pavonia*

### *Taonia atomaria* (Woodward)
J. G. Agardh

Frond 70–300mm long; membranous, translucent,
broadening out sharply from the base
and branching into wedge-shaped
growths; presence of reproductive
bodies and hairs imparts a striated
appearance.
Colour pale olive-green-brown
above; darker below.
Habitat usually on stones and
rocks down to 20m.
Distribution Mediterranean,
Atlantic north to south-west Britain
and Ireland, and English Channel.

*Taonia
atomaria*

*Polysiphonia lanosa*

**Note on the wracks** Members of the genera *Ascophyllum*, *Fucus* and *Pelvetia* are frequently found arranged in well-defined zones on the rocky shore where, together with the laminarians (see pages 32–33), they are often the dominant algae. Although there may be some local variations; a typical zonation pattern would be *Pelvetia canaliculata* on the upper shore; *Fucus spiralis* on the middle upper shore; *Ascophyllum nodosum* and/or *Fucus vesiculosus* around the middle part of the middle shore; *Fucus serratus* on the lower middle shore. N.B. *Fucus* spp. sometimes bear dense tufts of minute brown filaments on their fronds. These growths are the Brown Alga *Elachista fucicola* (Vell.) Aresch. reaching around 5mm.

### *Ascophyllum nodosum* (Linnaeus) Le Jolis
**Knotted Wrack**
Frond 300mm–1.500m long; tough, linear stem is rounded near the holdfast but flattened further along; main stem branches dichotomously several times and incorporates large air bladders spaced several millimetres apart; edge of main stem somewhat serrated, bearing many side-branches, many of which are slender and carry gold-yellow-olive reproductive bodies; lack of a midrib distinguishes these stems from those of *Fucus* (see below).
**Habitat** attached to rocks on the upper and middle shores.
**Distribution** Atlantic, English Channel and North Sea. N.B. this species is sometimes bearing the red tufts of *Polysiphonia* which are illustrated here growing on it.

### *Pelvetia canaliculata* (Linnaeus) Decaisne & Thuret
**Channelled Wrack**
Frond 50–150mm long; conspicuously grooved or channelled on one side; dichotomously branched; midrib absent; no air bladders, but tips swollen to form reproductive bodies.
**Habitat** on rocks on the upper shore where it forms a distinct zone.
**Distribution** Atlantic, English Channel and North Sea.

### *Fucus serratus* Linnaeus
**Toothed Wrack**
Frond about 600mm long but may be longer; short stalk or stipe develops into a tough, dichotomously branching frond; thick midrib; edge of frond conspicuously

*Ascophyllum nodosum*

*Pelvetia canaliculata*

*Fucus serratus*

serrated with teeth pointing in the direction of growth; minute hairs arranged in clusters along the frond; no air bladders.

**Colour** male plants may be more orange-brown than female plants.

**Habitat** attached to rocks on lower middle shore, often forming a distinct zone.

**Distribution** Atlantic, English Channel, North Sea and Baltic. N.B. often covered with epizoic animals, e.g. hydroids, bryozoans and spirorbid worms.

### *Fucus vesiculosus* Linnaeus
**Bladder Wrack**

**Frond** 150–1,000mm long; tough, strap-like, dichotomously branching; margin not serrated; usually with conspicuous air bladders arranged in groups of two or three; olive-brown-yellow reproductive bodies at the tip of the frond; conspicuous midrib.

**Habitat** on rocks on the middle shore except in very exposed places; forms a distinct zone.

**Distribution** Atlantic, English Channel, North Sea and Baltic. *Fucus vesiculosus* var. *linearis* ( =*F. evesiculosus*) (not illustrated) is a shorter form. Fronds lack air bladders and may appear damaged with smaller ones growing from the holdfast. Habitat: on rocks on exposed shores.

*Fucus versiculosus*

### *Fucus ceranoides* Linnaeus

**Frond** 300–600mm long (generally smaller than other species of Fucus); delicate, dichotomously branching; prominent midrib which may lack the lamina near the holdfast; pointed reproductive bodies grouped terminally on the branches in fan-like clusters.

**Habitat** growing on rocks and stones on the upper, middle and lower usually sheltered shores, usually where the salinity is reduced, e.g. in estuaries.

**Distribution** Atlantic, English Channel and North Sea.

### *Fucus spiralis* Linnaeus
**Spiral Wrack**

**Frond** 150–400mm long; tough, leathery; branches dichotomously and is usually twisted near the tips; margin not serrated; conspicuous midrib; conspicuous, rounded reproductive bodies borne on tips of branches surrounded by a flat flange; no air bladders.

*Fucus ceranoides*

**Habitat** on rocks on the upper shore below *Pelvetia* (opposite), but often absent from very exposed areas; may form a distinct zone.

**Distribution** Atlantic, English Channel and North Sea.

*Fucus spiralis*

### Fucus muscoides (Cotton) Feldm. and Magne
(Not illustrated)
**Frond** short, less than 40mm, narrow, 0.5–2mm wide.
**Colour** brown.
**Habitat** appearing as a mossy turf in the upper zone of very sheltered shores, common in inlets, sea lochs and salt marshes.
**Distribution** northern, Scotland, Scandinavia.

### Fucus distichus Linnaeus
Three subspecies of this variable form occur mainly on extreme northern coasts. See Hiscock, 1979.

### Bifurcaria bifurcata Ross ( =B. rotunda =B. tuberculata)
**Frond** 300–500mm long; main stem is rounded and unbranched for the lower quarter of its length, after which it bears alternately arranged branches; branches themselves fork irregularly or dichotomously, one sub-branch often larger than the other; swollen reproductive bodies may occur at branch tips; air bladders may develop along the thallus.
**Colour** brownish.
**Habitat** on rocks in pools and shallow water, never exposed to the air.
**Distribution** Atlantic north to Ireland, and English Channel.

*Bifurcaria bifurcata*

### Cystoseira baccata (Gmelin) Silva ( =C. fibrosa)
**Frond** 600–900mm long; main stem may be branched once or twice and bears many alternately arranged side-branches; the side-branches themselves bear many long, narrow branchlets which have a fine midrib and which taper at both the free end and at their bases; branchlets are often missing from the lower branches; air bladders present; bushy appearance overall.
**Colour** black when dry.
**Habitat** on rocks on the lower shore, in pools and shallow water.
**Distribution** Atlantic north to Ireland, and English Channel. N.B. several other species occur.

*Cystoseira baccata*

### Cystoseira tamariscifolia
(Hudson) Papenfuss
**( =C. ericoides)**
Frond 300–450mm long; cylindrical main stem may be branched a few times and bears many alternately arranged branchlets; branchlets themselves carry many short spines along their length as well as tufted reproductive bodies near their tips; air bladders may occur singly or in groups; overall appearance bushy.
**Colour** olive-brown; iridescent green-blue under water.
**Habitat** attached to rocks on the lower shore, in pools and shallow water.
**Distribution** Mediterranean, Atlantic and English Channel.

detail of reproductive bodies

*Cystoseira tamariscifolia*

### *Halidrys siliquosa* (Linnaeus) Lyngbye
**Sea-oak**
Frond 300mm–1.2m long; main stem bears regular and alternately arranged side-branches; plant somewhat compressed; air bladders resemble the seed pods of some terrestrial plants and are divided internally into about ten compartments.
**Holdfast** flattened and conical.
**Habitat** on rocks on the middle and lower shore and in shallow water.
**Distribution** Atlantic, English Channel and North Sea.

detail of reproductive bodies

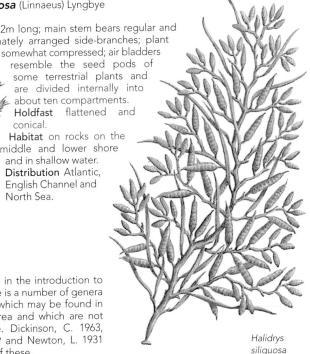

*Halidrys siliquosa*

As mentioned in the introduction to this section there is a number of genera of brown algae which may be found in the European area and which are not mentioned here. Dickinson, C. 1963, Hiscock, S. 1979 and Newton, L. 1931 describe many of these.

41

## CLASS RHODOPHYCEAE Red algae

Algae in which the chlorophyll is often masked by the red pigment phycoerythrin. They are exclusively multicellular and usually of small to moderate size. The red algae occur in temperate and warm waters at almost all sites along the shore and at various depths in the sea. See also Dixon P. S. & Irvine L. M. 1977 and Hiscock, S. 1986.

***Gelidium crinale*** (Turner) Lamouroux **( =G. aculeatum)**
**Frond** about 500mm long; compressed main stem bears branches arranged in pinnate fashion; these bear small, awl-shaped branchlets particularly at the apex; texture cartilaginous.
**Habitat** on sandy and rocky ground on the lower shore and in shallow water.
**Distribution** Mediterranean and Atlantic.

*Gelidium crinale*

***Gelidium latifolium*** (Greville) Bornet & Thuret
**Frond** about 80mm long; main branches flat and ribbon-like and bearing fine, thin branchlets; texture cartilaginous.
**Habitat** on rocks on the lower shore and in shallow water.
**Distribution** Atlantic north to English Channel, south-west Britain and Ireland.

*Gelidium latifolium*

***Gelidium sesquipedale*** (Clemente) Bornet & Thuret
**Frond** up to 200mm or more long; main stem slightly flattened bearing finer side-branches which taper towards their tips; texture more horny and less pliable than the two preceding species.
**Habitat** on rocks on the upper, middle and lower shores, in pools sometimes under other seaweeds and in deeper water.
**Distribution** Mediterranean, Atlantic as far as south-west England.
N.B. about three other species occur and some are difficult to identify.

*Gelidium sesquipedale*

**Nemalion helminthoides** (Velley) Batters
**( =N. elminthoides =N. multifidum)**
Frond 100–250mm long; stems worm-like and
branching, either just at the base or dichotomously
at various points; although the branches taper
towards their free ends, they are blunt at the tips
themselves, texture gelatinous or cartilaginous.
Holdfast minute and disc-like.
Colour brown-red.
Habitat middle shore on rocks and in pools, often in
fairly exposed places.
Distribution Atlantic (very rarely in North Sea
and Baltic).

*Nemalion
helminthoides*

*Scinaia
forcellata*

**Scinaia forcellata** (Turner) Bivona
Frond 50–250mm long; arranged cylindri-
cally, but occasionally flattened or compressed; dichotomous
branching repeated and regular with tips ending in a little
blunt fork; texture slimy.
Holdfast small and disc-like.
Colour pink-brown-red.
Habitat lower shore and shallow water.
Distribution Atlantic north to south-west Britain and Ireland.

**Pterocladia capillacea** (Gmelin)
Bornet & Thuret **( =P. pinnata)**
Frond 50–150mm long; generally larger
and stronger than most *Gelidium*
species; somewhat compressed and
with a tufted appearance; main stem may
not bear as many oppositely arranged
side-branches along its lower part as the
specimen illustrated; branches often taper
towards their bases and their free ends; frond
hollow; texture cartilaginous.
Habitat on rocks on the middle and lower shores
and in shallow water, often in calm places.
Distribution Mediterranean, Atlantic and
English Channel.

*Pterocladia
capillacea*

43

### Asparagopsis armata Harvey

This is the gametophyte generation of **Falkenbergia rufolanosa** (Harvey) Smith (see below). When the two plants were named it was not appreciated that they were heteromorphic generations of the same species. **Frond** 100–200mm long, slender, delicate, main stem bears irregularly placed branches; like the stem these branches are mainly covered with small, spirally distributed branchlets giving a tufted appearance; a few side-branches lack branchlets and bear alternate barbs or thorns; attached to substrate by a tangle of 'roots'.

**Habitat** in shady pools on the lower shore and also in shallow water.

**Distribution** Mediterranean and Atlantic, especially near south-west Britain (Scilly Isles) and west Britain (Isle of Man). N.B. distribution is often very local and this species, which was originally described in Australia, is thought to have been introduced to European waters in about 1925.

*Asparagopsis armata*

*Falkenbergia rufolanosa growing on another alga*

### Falkenbergia rufolanosa

(Harvey Smith)

This is the sporophyte generation of **Asparagopsis armata** Harvey (see above).

**Frond** consists of small tufts of fine, tangled filaments attached to various other seaweeds.

**Distribution** Much more widely distributed than *A. armata* and more common.

### Bonnemaisonia asparagoides

(Woodward) C. A. Agardh

This is the gametophyte generation of **Hymenoclonium serpens** (Crouan frat) Batters (not described in this book).

**Frond** 150–230mm long; rounded or compressed stem bearing alternate side-branches which are longest towards the base; the side-branches bear alternately arranged sub-branches and branchlets; branchlets are approximately the same length over the whole plant and arranged in the same plane.

**Holdfast** small and disc-like.

**Habitat** in shallow water.

**Distribution** Mediterranean, Atlantic, English Channel and North Sea.

*Bonnemaisonia asparagoides*

**_Furcellaria lumbricalis_** (Hudson) Lamouroux
Frond 100–200mm long; stiff, cylindrical, glossy
stem branches regularly and equally; succeeding
branches become regularly shorter towards the
tips; in Winter the tips bear pod-like reproductive
bodies, but these may be lacking at other times
of the year so that the tips then appear more
tapered.
**Holdfast** is a mass of branching rootlets up
to 25mm across.
**Habitat** on rocks on the lower shore and
in shallow water.
**Distribution** Atlantic, English Channel,
North Sea and Baltic. N.B. the genus
_Polyides_ (see page 51) is very similar, but
may be differentiated from _F. lumbricalis_
by its rounded, discoidal holdfast.

**_Halarachnion ligulatum_** (Woodward) Kützing
Frond 100–300mm long; strap-like stems, dichotomously
branching, may bear many side-branches sometimes arranged
irregularly or opposite; side-branches much narrower than
the stem and often terminating in a notch; texture soft and
gelatinous.
**Colour** pink-red.
**Habitat** on rocks and shells in shallow water; may be washed up.
**Distribution** Mediterranean, Atlantic and English Channel.

_Furcellaria
lumbricalis_

**_Catenella caespitosa_** (Withering) Dixon & L. Irvine
**( =_C. opuntia_)**
Frond about 30mm long; moss-like growth of
creeping fibres from which tiny, irregularly
branching fronds arise; may grow in clumps
about 50mm across.
**Habitat** on upper shore on rocks and in
crevices, sometimes among the holdfasts of
_Fucus spiralis_ and _Pelvetia canaliculata_ (see
pages 38–39).
**Distribution** Atlantic, English Channel and
North Sea. N.B: superficially like _Gelidium_
species (see page 42).

_Halarachnion
ligulatum_

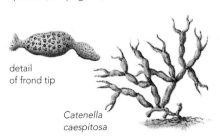

detail
of frond tip

_Catenella
caespitosa_

45

### *Calliblepharis ciliata* (Hudson) Kützing

**Frond** 150–300mm long; flat, strap-like main stem increases in width and branches irregularly; stem borne on a short, rounded stipe; short, awl-shaped side-branches occur all round the margin of the main stems and on their upper and lower surfaces.

**Holdfast** consists of branching 'roots'.

**Habitat** on rocks and in pools on middle and lower shores and in shallow water.

**Distribution** Atlantic, north to south-west Britain and Ireland, and English Channel.

*Calliblepharis ciliata*

*Cystoclonium purpureum*

### *Cystoclonium purpureum* (Hudson) Batters

**Frond** 150–600mm long; rounded, succulent stems bear many irregularly disposed branches which themselves bear many branchlets; branches and branchlets taper towards their bases, often more so than in the specimen illustrated; branchlets noticeably finer than branches; rounded reproductive bodies may occur towards the tips of the branchlets.

**Holdfast** root-like.

**Habitat** on rocks and other seaweeds on the lower share and in shallow water.

**Distribution** Atlantic, English Channel, North Sea and Baltic.

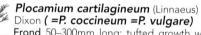

### *Plocamium cartilagineum* (Linnaeus) Dixon ( *=P. coccineum =P. vulgare)*

**Frond** 50–300mm long; tufted growth with strong main stems tending to have more irregularly arranged branches in the upper region and fewer or none at the base; branches bear branchlets whose terminal subdivisions all occur on the same side (see inset illustration); almost globular reproductive bodies may occur over the plant.

**Habitat** on rocks in shallow water; may be washed up.

**Distribution** Mediterranean, Atlantic, English Channel and North Sea.

*Plocamium cartilagineum*

### *Phyllophora crispa* (Hudson) Dixon ( *= P. epiphylla =P. rubens)*

**Frond** up to 250mm long; flat, ribbon-like stem borne on a very short, compressed or cylindrical stipe; branching may be dichotomous; branches generally have blunt tips and often arise from face at blade; rigid, crisp texture.

**Holdfast** very small and disc-like, but may coalesce with those of neighbours.

**Habitat** usually attached to vertical rocks and pool sides on lower shore and in shallow water.

**Distribution** Mediterranean and Atlantic. N.B. about three other species occur.

*Phyllophora crispa*

### *Hypnea musciformis* (Wulfen) Lamouroux
**Frond** 100–300mm long; upright stem bears a number of irregularly disposed, almost pinnate branches.
**Colour** black-red-green.
**Habitat** almost always growing entangled with other seaweed, generally in sheltered places.
**Distribution** Mediterranean and Atlantic.

### *Ahnfeltia plicata* (Hudson) Fries
**Frond** 70–150mm long; forming stiff, springy tufts about 100mm wide with a texture reminiscent of fine wire.
**Holdfast** is a thin, encrusting structure up to 20mm in diameter.
**Habitat** in pools and attached to rocks on the middle and lower shore.
**Distribution** Atlantic, English Channel and North Sea.

### *Phyllophora pseudoceranoides* (S. G. Gmelin) Newroth
& A. R. A. Taylor
**Frond** 70–200mm long; generally fan-like, borne on a fairly long, cylindrical stipe or stalk, and branching into flattened, wedge-shaped 'leaves'; branching dichotomous; branches terminate in a notch.
**Holdfast** small and disc-like.
**Colour** usually brown-purple-red.
**Habitat** usually attached to horizontal rocks, on lower shore and occasionally in shallow water.
**Distribution** Atlantic, English Channel, North Sea and Baltic.

### *Gracilaria verrucosa* (Hudson) Papenfuss
### ( =G. confervoides)
**Frond** 70–500mm long; stringy stem grows up with a number of irregular branches which in turn may carry many slender branchlets; branchlets taper at their bases and at their free ends; reproductive bodies which resemble small, wart-like protuberances may be scattered over the plant; several stems may arise from the same fleshy discoidal holdfast; frond may be partly prostrate and one of the few seaweeds to anchor in sand.
**Habitat** on rocks, gravel and sand on the middle shore and down to shallow water.
**Distribution** Mediterranean, Atlantic and English Channel.

*Hypnea
musciformis*

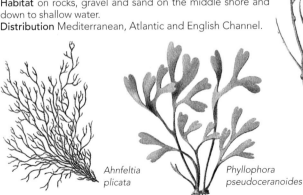

*Ahnfeltia
plicata*

*Phyllophora
pseudoceranoides*

*Gracilaria
verrucosa*

### *Chondrus crispus* Stackhouse
**Irish Moss or Carragheen**
(when collected with *Gigartina*)
Frond 70–150mm long; basal part of stem is narrow and unbranched; upper stem branches dichotomously, tips often iridescent; stem not channelled; branches form wedge-shaped segments, somewhat variable in appearance with narrow and broader forms existing, texture cartilaginous.
**Holdfast** discoidal.
**Colour** red-purple may turn green in strong light.
**Habitat** on stones and rocks on the lower shore and in pools, also in shallow water.
**Distribution** Atlantic, English Channel, North Sea and rarely in the Baltic. N.B. may be confused with *Mastocarpus stellata* (see below) and *Gymnogongrus crenulatus* (see opposite).

*Chondrus crispus 2 forms*

### *Mastocarpus stellata* (Stackhouse) Batters
**Frond** 100–200mm long, but is often shorter than *Chondrus crispus* (above); tufted appearance; flattish and somewhat channelled or inrolled; narrow at base but becoming broader and more strap-like towards the apex; dichotomously branched; fertile specimens may be covered with small pimples which, in addition to the channelling, further help to distinguish it from *Chondrus crispus*.
**Habitat** on rocks on extreme lower shore.
**Distribution** Atlantic, English Channel and the North Sea.

### *Gigartina acicularis*
(Wulfen) Lamouroux
(Not illustrated)
**Frond** 40–100mm; not similar to *M. stellata* (above), frond not channelled but rather more rounded, and slender.
**Colour** purple-red.
**Distribution** Mediterranean and Atlantic north to south-west Britain and Ireland where it is rare. N.B. about three other species occur.

*Mastocarpus stellata*

*Hildenbrandia rubra*

### *Hildenbrandia rubra*
(Sommerfield) Meneghini
**Thallus** maybe extensive, sometimes consisting of a large patch of thin tissue closely adhering to a stone or rock; with care it may be peeled off; loses its sheen when dry; spores born in pits, visible with hand lens.
**Colour** pink or brown-red.

*Gigartina acicularis*

Habitat on all levels of the shore, including splash zone, often situated so that it keeps moist when the tide is out.
Distribution Mediterranean, Atlantic, English Channel, North Sea and Baltic.

### *Gymnogongrus crenulatus* (Turner) J. Agardh
Frond reaching 100mm and somewhat flattened which expands into branches reaching 40mm broad. Branch tips typically square ended; never iridescent; dark reproductive 'lumps' occur on surface.
Colour dark brownish red, no purple hue.
Habitat in pools on shore and in shallow water, sometimes near sand.
Distribution Atlantic, English Channel.

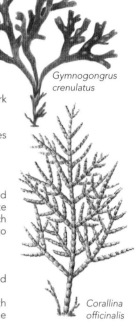

*Gymnogongrus crenulatus*

### *Corallina officinalis* Linnaeus
Frond 50–120mm long; main stem bears branches arranged exactly opposite; branches in their turn bear opposite branchlets; plant consists of number of calcified segments which are longer than broad, and which are linked by pliable joints to form the frond; terminal reproductive bodies lack 'horns'.
Holdfast chalky and encrusting.
Colour varies from purple-red-pink to yellow-white.
Habitat on rocks and pools at all levels of the middle shore and down to shallow water.
Distribution Mediterranean, Atlantic, English Channel, North Sea and rarely in the Baltic. N.B. avoid confusion with the thecate hydroids (see page 81). Several other species occur.

*Corallina officinalis*

### *Corallina elongata* Ellis & Solander
### ( =*C. mediterranea*)
Frond up to 80mm long; similar to *C. officinalis* (above), but branches from base giving a tufted appearance, branches bear pinnate branchlets; stem segments ovoid or slightly triangular; terminal reproductive bodies bear horns.
Habitat attached to rocks in pools and in shallow water.
Distribution Mediterranean, and Atlantic north to English Channel.

### *Jania rubens* (Linnaeus) Lamouroux
### ( =*Corallina rubens*)
Frond 20–50mm long; chalky and jointed as in *Corallina* (above) but branching is dichotomous and not opposite; often growing in dense tufts; in Spring there are conspicuous, rounded reproductive bodies.
Holdfast minute and disc-like.
Colour rose-red.
Habitat growing attached to other seaweeds especially on *Cladostephus* sp. (see page 36).
Distribution Mediterranean, Atlantic, English Channel and rarely in North Sea.

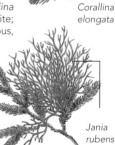

*Corallina elongata*

*Jania rubens*

49

*Lithophyllum incrustans*

### *Lithophyllum incrustans* Philippi

**Thallus** consists of a patch of chalky tissue up to 40mm thick; growths may be irregular in outline and smooth or rather bumpy in appearance, sometimes lying on top of each other; adheres strongly to the substrate and may embrace small shells, etc.; the margin of the growth tends to be less differentiated than the rest in terms of colour or texture (c.f. *Phymatolithon calcareum* below).

**Colour** mauve-purple-red-yellow, being darker in shady places.

**Habitat** in rock pools on middle and lower shores; appears to prefer exposed places.

**Distribution** Mediterranean, Atlantic, English Channel and North Sea. N.B. about six other species occur. *Hildenbrandia prototypus* is extremely thin and is not calcareous (see pages 48–49).

*Fosliella farinosa on Zostera leaf*

### *Fosliella farinosa* (Lamouroux) Howe

**Thallus** circular or irregular, forming encrusting growths on the fronds of other seaweeds, *Zostera* (see pages 64–65), hydroids and worm tubes.

**Colour** pink.

**Habitat** attached to hosts on the extreme lower shore and in shallow water.

**Distribution** Mediterranean, Atlantic, English Channel and North Sea. N.B. about twenty other species occur.

*Phymatolithon calcareum*

### *Phymatolithon calcareum* (Pallas) Adey & McKibbon
**Maerl**

**Thallus** somewhat similar to *Lithophyllum incrustans* (above) when young, and distinguished by slightly thicker margins; older specimens become somewhat erect with nodular branches and are reminiscent of red 'coral' (not illustrated); growths may reach 80mm across.

**Colour** violet-red.

**Habitat** unattached, or occasionally encrusting pebbles on extreme lower shore and shallow water.

**Distribution** Mediterranean, Atlantic, English Channel and North Sea. N.B. about fifty other species occur.

*Audouinella floridula*

### *Audouinella floridula* (Dillw.) Woelkerling
**( =Rhodochorton species)**

A difficult group of filamentous algae. Smooth, apparently structureless masses of crimson threads lying in accumulated sand, often covering rocks. Filaments themselves velvety, minute in diameter. Plants bind sand and stabilise it.

**Habitat** middle shore, usually in composite rock/sand.

**Distribution** Atlantic, English Channel and North Sea.

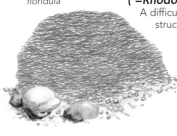

### Polyides rotundus (Hudson) Greville ( =P. caprinus)

Frond 100–200mm long; rounded stem usually free of branches near the base, but dividing dichotomously about one-third of the way up the stem.

Holdfast disclike (not root-like as in *Furcellaria*, see page 45).

Colour dull red, turning black when dry.

Habitat on rocks and stones and in pools on lower shore and down to 20m.

Distribution Atlantic, English Channel and North Sea.

*Polyides rotundus*

*Dilsea carnosa*

### Dilsea carnosa (Schmidel) Kuntze
Frond 130–300mm long; short, rounded stipe gives way to a flattened, leaf-like frond; texture is smooth; the fronds may be split in older specimens; several fronds may develop from the discoidal holdfast.
Colour usually dark red.
Habitat attached to rocks on extreme lower shore and in shallow water.
Distribution Atlantic, English Channel and North Sea.

### Grateloupia filicina (Lamouroux) C. A. Agardh (includes G. minima)
Frond 50–120mm long; tufted, compressed growth; main stem tapers at free end and at base, and bears alternate or opposite branches; branches carry alternate or opposite branchlets.
Holdfast discoidal.
Habitat on stones and rocks on middle shore and down to shallow water, usually near freshwater outfalls.
Distribution Mediterranean, Atlantic north to west Britain and Ireland, and English Channel.

*Grateloupia filicina*

### Dumontia incrassata (O. F. Müller) Lamouroux
Frond 20–500mm long; rounded stem is unbranched in lower regions, but higher up bears alternately or irregularly arranged branches; branches rounded and tapering towards bases and free ends; branchlets rarely present; in young specimens the stem and branches are solid, but they become hollow with age.
Colour dull red in shady places, but becoming yellow-green-brown in strongly illuminated positions.
Habitat on rocks and pebbles or on other seaweeds, often in pools on middle shore.
Distribution Atlantic, English Channel and North Sea.

*Dumontia incrassata*

51

### Callophyllis laciniata (Hudson) Kützing

Frond about 80mm long; thickish, flat and quickly broadening out from the very short stipe; split up into a number of wedge-shaped sections by a form of dichotomous branching; branch tips blunt; reproductive bodies may occur in tiny leaflets arranged around the periphery of frond in Summer, or as minute dots over the entire surface.

Holdfast small and disc-like.

Colour opaque pink-red.

Habitat on kelp stipes, rocks and stones in shallow water and on kelp holdfasts; often washed up.

Distribution Atlantic, English Channel and North Sea.

Callophyllis
laciniata

### Gastroclonium ovatum (Hudson) Papenfuss

Frond 50–150mm long; cartilaginous, round stem branches dichotomously or alternately; upper part of plant bears small pip-like 'leaves'.

Holdfast small and disc-like.

Habitat on rocks an the lower shore and in pools.

Distribution Atlantic north to Scotland, and English Channel.

Gastroclonium
ovatum

### Lomentaria articulata (Hudson) Lyngbye

Frond 50–250mm long; consists of a hollow stem and branches which are repeatedly constricted to give a jointed or beaded effect; branches only occur at the joint between two 'beads'; main stem branches dichotomously from the base and there after the pattern may be opposite; very small holdfast and attachment runners from the stem; internally gelatinous.

Colour dull purple-bright red; shiny, transparent.

Habitat attached to rocks and other seaweeds on the upper, middle and lower shores.

Distribution Atlantic, English Channel and North Sea.

Lomentaria
articulata

Lomentaria
clavellosa

### Lomentaria clavellosa (Turner) Gaillon

Frond 70–400mm long; undivided, rounded main stem bears opposite or alternate pinnate branches; rounded branches and branchlets taper at bases and free ends; texture pliable and gelatinous.

Colour bright red-pink.

Habitat on rocks and other seaweeds on lower shore and in shallow water.

Distribution Atlantic, English Channel and North Sea.

## Antithamnion cruciatum
(C. A. Agardh) Nägeli

**Frond** 25–50mm long; main stem bears alternate branches and branchlets; tips of branches tufted; under the microscope consists of many single, oblong cells joined end to end, each cell bearing 2 opposite branchlets.

**Habitat** on rocks in muddy places on lower shore.

*Antithamnion cruciatum*

**Distribution** Mediterranean, Atlantic and English Channel. N.B. about four other species occur.

## Callithamnion arbuscula (Dillwyn) Lyngbye
**Frond** 50–150mm long; bushy growth with short stipe; main stem zig-zags, bearing alternate branches and branchlets; apices of branches tufted.
**Holdfast** small, stout and discoidal.
**Habitat** on steep rocks on lower shore; often exposed.
**Distribution** Atlantic north from England and northern Ireland, and North Sea. N.B a number of other species occur.

*Callithamnion arbuscula*

## Callithamnion corymbosum (Smith) Lyngbye
**Frond** 20–70mm long; main stem bears alternately arranged branches which carry delicate, filamentous, alternate branchlets.
**Habitat** on rocks and seaweed on lower shore and in shallow and deeper water.
**Distribution** Mediterranean, Atlantic, North Sea and Baltic. N.B about ten other species occur.

*Callithamnion corymbosum*

## Rhodymenia pseudopalmata (Lamouroux) Silva ( =R. palmetta)
**Frond** 40–100mm long; short stipe up to 25mm long, broadening into a flattened, fan-like frond with dichotomous branching; tough, rigid; unlike *P. palmata* (below) no peripheral 'leaflets'.
**Holdfast** disclike.
**Habitat** on rocks and laminarian stipes on lower shore and in shallow water where they and related species may be abundant towards the deep limit of algae, especially in south-west Britain.
**Distribution** Atlantic north to English Channel, and south-west Britain.

*Rhodymenia pseudopalmata*

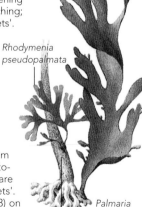

## Palmaria palmata (Linnaeus) O. Kuntze
**Dulse**
**Frond** 100–300mm long; no stipe; gradually expands from the wide, disc-like holdfast to form a flattened, dichotomous 'fan'; younger parts are delicate, older parts are tougher and darker and may bear small peripheral 'leaflets'.
**Habitat** on rocks and laminarian stipes (see pages 32–33) on middle and lower shores and in shallow water.
**Distribution** Atlantic, English Channel and North Sea.

stalk of laminarian

*Palmaria palmata*

53

### Ceramium rubrum (Hudson) C. A. Agardh

**Frond** 20–300mm long; main stem branches dichotomously but unevenly; side-branches arranged likewise; terminal branchlets have tips which often point inwards like forceps; close inspection reveals a banded pigmentation effect on the sterns and branches; overall appearance bushy; texture cartilaginous.

**Holdfast** minute and cone-like.

**Colour** variable; deep red-brown-yellowish.

**Habitat** on rocks and other seaweed on upper middle and lower shores and in shallow and deeper water.

**Distribution** Mediterranean, Atlantic, English Channel, North Sea and Baltic. N.B. about twenty-three other species occur. This is a very variable species.

detail of terminal branchlet

*Ceramium rubrum*

### Griffithsia flosculosa (Ellis) Batters

**Frond** 70–200mm long; delicate, filamentous growth; main stern branches dichotomously but unevenly; filaments gradually taper towards the apex; reproductive bodies like minute spheres borne on small stalks may occur all over the filaments; texture when in the sea is stiff, but it quickly wilts when removed from water.

**Holdfast** a tangle of 'rootlets'.

**Colour** quickly lost if this species is placed in fresh water.

**Habitat** attached to rocks in pools on the lower shore and in shallow water.

**Distribution** Atlantic, English Channel and North Sea, N.B. about three other species occur.

### Plumaria elegans (Bonnemaison) Schmitz

**Frond** 50–100mm long, main stem branches dichotomously or alternately; branches usually bear alternate or opposite branchlets.

**Holdfast** small and fibrous.

**Colour** brown-purple.

**Habitat** on shaded rocks and under overhangs, often hanging down in dingy clusters, on middle and lower shore.

**Distribution** Atlantic, English Channel and North Sea. N.B. confusion with *Ptilota plumosa* (see opposite) can be avoided by examining the sub-terminal branchlets. In *Plumaria elegans* they are translucent. Also, in *Plumaria elegans* the main stem is

*Griffithsia flosculosa*

*Plumaria elegans*

covered with minute branchlets whereas these are lacking in *Ptilota plumosa*.

### *Ptilota plumosa* (Hudson) C. A. Agardh

**Frond** 100–300mm long; main stem rather stiff and irregularly branched in one plane; pinnate branchlets; subterminal branchlets are opaque and the main stem is smooth and free from minute branchlets (c.f. *Plumaria elegans* above).
**Holdfast** a disc of matted fibres.
**Habitat** generally on stipes of laminarians (see pages 32–33) on extreme lower shore and in shallow water.
**Distribution** Atlantic north from England and Ireland and North Sea.

### *Halurus equisetifolius* (Lightfoot) Kützing
**Sea-tail**

**Frond** 70–220mm long; main stem branches irregularly in all planes around the vertical axis; the stiff stem and branches bear many whorls of small filaments each about 20mm long, thus giving the plant a bottle-brush appearance.
**Colour** dark red.
**Habitat** on rocks on the middle and lower shores, sometimes in pools and shallow water.
**Distribution** Atlantic and English Channel.

### *Sphondylothamnion multifidum* (Hudson) Nägeli

**Frond** 100–200mm long; main stem usually undivided and branching in one or two planes; branches arranged opposite or alternately; similar arrangement of sub-branches; whole plant covered in whorls of branchlets but these are much more separated than in *Halurus equiseti-folius* (see above); texture crisp when fresh, goes soft when collected.
**Holdfast** well-developed and 'root-like'.
**Colour** pink-red.
**Habitat** on rocks on vertical sides of pools on lower shore and down to shallow water.
**Distribution** Atlantic and English Channel.

*Ptilota plumosa*

*detail of branch*

*Halurus equisetifolius*

*Sphondylothamnion multifidum*

55

*Apoglossum
ruscifolium*

### Apoglossum ruscifolium (Turner) J. G. Agardh

**Frond** 60–100mm long, broad, leafy main stem with midrib terminates in a rounded tip; edges waved; main stem bears alternately arranged branches and sub-branches which develop from the midrib; under a hand lens fresh specimens show wide-angled veins; diverging from the midrib; short stipe.

**Habitat** on rocks and in pools as well as on laminarians on the middle and lower shore.

**Distribution** Atlantic, English Channel and North Sea. N.B. avoid confusion with *Hypoglossum woodwardii* which occurs in similar habitats (see below).

### Cryptopleura ramosa (Hudson) Newton

**Frond** 100–200mm long; entwined, bushy growth generally as wide as it is long; frond bare and has strong veins, the lowermost region may lack the 'leaf'; edges of leafy stem waved and sometimes with small peripheral growths; stem bears irregular dichotomous branches which are somewhat wedge-shaped with rounded tips.

**Holdfast** small and disc-like.

**Colour** brown-red-purple; somewhat iridescent.

**Habitat** on rocks and laminarian stipes on the lower shore and in shallow water.

*Cryptopleura
ramosa*

**Distribution** Atlantic, English Channel and North Sea.

*Delesseria
sanguinea*

### Delesseria sanguinea (Hudson) Lamouroux

**Frond** up to 400mm over all; consists of a rounded, branching stalk or stipe (up to 150mm) and 'leaves' (100–250mm); 'leaves' are oval – lance-shaped, pointed at tip when young and rounded when older; 'leaves' show conspicuous midrib and 'veins'.

**Habitat** attached to rocks or laminarians in deep pools on lower shore in shade, or in shallow water.

**Distribution** Atlantic, English Channel, North Sea and rarely in the Baltic. N.B. avoid confusion with *Phycodrys rubens* (see right).

### Hypoglossum woodwardii Kützing

**Frond** 50–200mm long; main stem resembles a narrow leaf and bears alternately arranged branches and sub-branches which develop from the conspicuous midrib; under a hand lens, indistinct transverse veins may be seen; main stem and branches terminate with sharp points; several main stems may arise from disc-like holdfast.

*Hypoglossum
woodwardii*

Habitat on rocks in pools and on laminarians on the middle and lower shores.
Distribution Atlantic, English Channel and North Sea. N.B. avoid confusion with *Apoglossum ruscifolium* (see opposite).

### *Nitophyllum punctatum* (Stackhouse) Greville

Frond 100–500mm long; broad, wedge-shaped, membranous frond grows directly from the small, disc-like holdfast; one or more may develop from each attachment point; branching is fairly regular and dichotomous; the presence of many small, terminal branchlets may make the margins look frilly; no 'veins', texture delicate.
Colour red-pink.
Habitat on various seaweeds in pools on lower shore and in shallow water.

*Nitophyllum punctatum*

Distribution Mediterranean, Atlantic and English Channel. N.B. avoid confusion with *Cryptopleura ramosa* (see page 56).

### *Membranoptera alata*

(Hudson) Stackhouse
Frond 100–200mm long; flat main stem is like a narrow leaf with a conspicuous midrib, and under a hand lens shows distinct, fine 'veins'; irregular dichotomous branching; branches tapering towards the tips which are often notched.
Holdfast small and discoidal.
Habitat on rocks or on other red and brown seaweeds in shaded pools on middle and lower shore, and in shallow water often on kelp holdfasts.
Distribution Atlantic, English Channel, North Sea and Baltic. N.B. can be distinguished from *Apoglossum* and *Hypoglossum* (see page 56) by the form of branching.

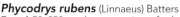

*Membranoptera alata*

### *Phycodrys rubens* (Linnaeus) Batters

Frond 50–250mm long; consists of a branching stem and 'leaves'; these are much indented and 'veined' and resemble oak leaves; generally more branched and with shorter stems than *Delesseria* (opposite); colour is paler also.
Holdfast small and disc-like.
Habitat in shady places on rocks and on stipes of laminarians on lower shore; often washed up.
Distribution Atlantic, English Channel, North Sea and west Baltic.

*Phycodrys rubens*

### Dasya hutchinsiae Harvey ( =D. arbuscula)

Frond 70–100mm long; main stem bears alternately arranged side-branches; these bear fine, filamentous branchlets; small, flask-shaped reproductive bodies may be carried on stalks by branchlets.

Holdfast consists of 'rootlets'.

Colour brown-crimson.

Habitat on rocks on the lower shore and in shallow water.

Distribution Mediterranean, and Atlantic northwards to west coast of Ireland.

*Dasya hutchinisiae*

*Heterosiphonia plumosa*

### Heterosiphonia plumosa (Ellis) Batters

Frond 150–200mm long; flattened, somewhat feathery in appearance; main stem tapers towards free end; branches arranged alternately or irregularly, bearing sub-branches and branchlets.

Colour red-deep crimson.

Habitat on rocks or other seaweeds on extreme lower shore and in shallow water; may be washed up.

Distribution Atlantic north to Ireland, English Channel and North Sea.

### Bostrychia scorpioides

(Hudson) Montagne

Frond 50–100mm long; plant tufted or tangled, main stem irregularly or alternately branched; fine alternate branchlets; branches very often coiled towards the tips.

Holdfast absent.

Colour brownish.

Habitat generally among the roots and stems of other plants (especially angiosperms) on the upper shore; very often in estuaries and salt marshes.

Distribution Atlantic north to England and Ireland, English Channel and North Sea.

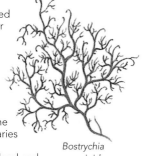

*Bostrychia scorpioides*

### *Laurencia pinnatifida* (Hudson) Lamouroux
**Pepper Dulse**
**Frond** 20–300mm long, flattened, well-developed main stem alternately branched, branches subdivided into smaller branchlets; texture cartilaginous.
**Holdfast** disc-like with 'rootlets'.
**Colour** variable, depending on position on the shore, but often purple- brownish, may be green-yellow in the middle tide range.
**Habitat** on rocks and in crevices from the middle shore down to shallow water.
**Distribution** Mediterranean, Atlantic, English Channel and North Sea. N.B. bleaches in strong light.

*Laurencia pinnatifida*

### *Laurencia obtusa* (Hudson) Lamouroux
**Frond** 70–150mm long; forms thick, globular growths; round main stems bears alternate or opposite branches which may be arranged spirally on the stem; branchlets similarly arranged; branches and branchlets become smaller towards the tips, cartilaginous texture.
**Colour** variable; purple-pink-yellow.
**Holdfast** small and disc-like, sometimes with 'rootlets'.
**Habitat** usually on other seaweeds in pools on lower shore and in shallow water.
**Distribution** Mediterranean, Atlantic, English Channel and North Sea. N.B. bleaches in strong light.

*Laurencia obtusa*

### *Odonthalia dentata* (Linnaeus) Lyngbye
**Frond** 70–300mm long; flattened plant with tufted appearance; midrib can be made out near the base; main stem bears alternate branches; sub-branches and branchlets alternately arranged; branches and branchlets have a toothed appearance.
**Habitat** on rocks and, rarely, on other seaweeds such as laminarians on the lower shore and in shallow water.
**Distribution** Atlantic north from England and Ireland, and North Sea.

*Odonthalia dentata*

59

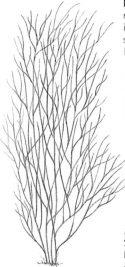

Note on the genus *Polysiphonia* This large genus contains many poorly defined species. About thirty may be encountered in the European area of which the following four have been selected. A more complete coverage is provided by Newton, L. 1931.

### *Polysiphonia urceolata* (Dillwyn) Greville

Frond 50–250mm long; filamentous with a tufted appearance; main stems bear alternate branches in all planes around the long axis; several stems may develop from the creeping 'roots' which form the holdfast; may be found with characteristic urn-shaped, reproductive bodies.

Colour purple-red-brown.

Habitat on stones and other seaweeds and shells on the lower shore and down to 20m.

Distribution Mediterranean, Atlantic, English Channel, North Sea and Baltic.

*Polysiphonia
urceolata*

### *Polysiphonia elongata* (Hudson) Sprengel

Frond 150–300mm long; well-defined main stem; bushy appearance; branches develop about 20mm from base and are arranged alternately around the stem; sub-branches bear fine, clustered, terminal branchlets especially in Spring; texture gelatinous. Holdfast small, consists of 'rootlets'.

Colour dark red-yellowish. Habitat on rocks, stones and shells as well as on some larger seaweeds on lower shore and in shallow water.

Distribution Mediterranean, Atlantic, English Channel, North Sea and Baltic.

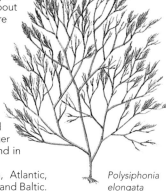

*Polysiphonia
elongata*

### *Polysiphonia lanosa* (Linnaeus) Tandy *( =P. fastigiata)*

Frond up to 80mm long; a strong, filamentous plant growing in tufts on *Ascophyllum nodosum* and, rarely, on *Fucus* (see pages 38–40); main stem bears alternate or dichotomous branches in all planes around the long axis; branches become shorter from the base; sub-branches similarly arranged; creeping 'rootlets' invade the host's tissue.

Habitat on seaweeds.

Distribution frequent where-ever *Ascophyllum* is to be found: Atlantic, English Channel and North Sea. N.B. this species may in turn be affected by the small, cushion-like algal parasite *Choreocolax* (not featured in this book).

*Polysiphonia lanosa
on Ascophyllum
nodosum*

### Polysiphonia nigrescens (Hudson) Greville

**Frond** 70–300mm long; growing in thick clusters; main stem may be much divided and bears alternate or irregular branches which in turn bear numbers of soft branchlets; some of the lower branches may be lost in older plants.
**Holdfast** of branching 'rootlets'.
**Colour** purple-brown; older plants become blackish.
**Habitat** on rocks, stones and other seaweeds in pools on the middle shore.
**Distribution** Atlantic, English Channel, North Sea and Baltic.

*Polysiphonia nigrescens*

### Rhodomela confervoides (Hudson) Silva ( =R. subfusca)

**Frond** 70–300mm long; cylindrical main stem bears spirally arranged branches which are shorter from the base up; the branches may be divided and covered with fine, filamentous branchlets; plants which have over-wintered may show areas of softer growth in Spring; texture generally cartilaginous.
**Holdfast** disc-like.
**Colour** brown-red.
**Habitat** on rocks and seaweeds from lower shore down to about 20m.
**Distribution** Atlantic, English Channel, North Sea and Baltic. N.B. close examination with a hand lens reveals the lack of 'joints' in the tips of the branches which distinguishes this species from *Polysiphonia*.

*Rhodomela confervoides*

### Porphyra umbilicalis (Linnaeus) J. G. Agardh

**Frond** 50–200mm long; irregular, gelatinous, membra-nous growth arranged in 'leaves', usually attached at one point.
**Colour** red-purple-green, becoming black when specimen is dry.
**Habitat** on stones and rocks especially when covered with sand.
**Distribution** Mediterranean, Atlantic and North Sea. N.B. about five other species occur.

### Myriogramme bonnemaisonii

**Frond** rose to pale pink, fan shaped with divided edges, dichotomously branched; frond thickens near base; reaches 150mm in length.
**Habitat** lower shore and into deep water, attached to kelp stipes, other algae and rocks.
**Distribution** Atlantic, English Channel.

*Porphyra umbilicalis*

*Myriogramme bonnemaisonii*

Lichens have been less studied than the algae. They exist as dual plants, partly composed of fungal tissue and partly composed of algal tissue. Formerly they were classified as a discrete group, but recent work has shown that they should be classified with the fungi. Their life processes resemble both those of the algae and the fungi, but they also display particular properties of their own. Many botanical text books provide an excellent account of their structure and function. Here it may be said that the lichen thallus usually consists of flat or leafy outgrowths which encrust other plants, rocks or shells. The tissue may be brittle or soft, rough or smooth, tufted, branched or flat. Many lichens have fruiting structures which may assist with their identification. The precise identification of these organisms is in many cases a specialized operation, however, and reference should be made to Duncan, U. K. 1959, Dobson, F. 1979, Dobson, F. 1997 or Ferry, B. W. and Sheard, J. W. 1969. Unless otherwise stated the lichens described below will be found on the upper shore and on the rocks above it.

*Roccella fuciformis*

### *Roccella fuciformis* (Linnaeus) Candolle
**Thallus** strap-like, pendent; 50–200mm long; groups of flat, disc-like, fruiting bodies full of dusty, black spores.
**Distribution** Atlantic north to south-west England and Ireland.

### *Ramalina siliquosa* (Hudson) A. L. Smith
**Thallus** strap-like, erect; 20–50mm long; surface smooth and glossy; disc-like fruiting bodies coloured white-brown near tips of thallus.
**Distribution** Atlantic, English Channel and North Sea.

### *Ochrolechia parella* (Linnaeus) Massalongo
**Thallus** encrusting; forms rounded patches 30–100mm in diameter; conspicuous, saucer-like fruiting bodies with grey, dusty spores.
**Distribution** Atlantic, English Channel and North Sea.

*Ramalina siliquosa*

### *Lichina pygmaea* (Lightfoot) C. A. Agardh
**Thallus** tufted, branching; about 10mm high resembling a minute seaweed.
**Habitat** upper and middle shores.
**Distribution** Atlantic, English Channel and North Sea. N.B. a related species occurs in the Mediterranean.

### *Verrucaria mucosa* Wahlenberg
**Thallus** smooth, closely attached to rocks; irregular shape up to 300mm across; texture may be jelly-like.
**Habitat** middle shore.
**Distribution** Atlantic and English Channel. N.B. a related species occurs in the Mediterranean.

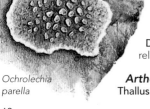

*Ochrolechia parella*

### *Arthopyrenia sublittoralis* (Leighton) Arnott
**Thallus** growing in minute cavities in shells of barnacles,

etc; a hand lens will reveal the lichen in its pit.

**Habitat** upper and middle shores.

**Distribution** Atlantic and English Channel, rarely in North Sea.

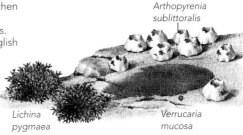

*Arthopyrenia sublittoralis*

### Xanthoria parietina

(Linnaeus) T. A. Fries

**Thallus** rough and leafy; 30–100mm across.

**Habitat** often form a belt just below *Ramalina* (opposite) on extreme upper shore.

**Distribution** Atlantic, English Channel and North Sea.

*Lichina pygmaea*

*Verrucaria mucosa*

### Caloplaca marina Weddell

**Thallus** flattish and encrusting with scattered, coarse granules forming patches up to 100mm across; not leafy.

**Distribution** Atlantic, English Channel and North Sea.

N.B. a related species occurs in the Mediterranean.

*Xanthoria parietina*

### Lecanora atra (Hudson) Acharius

**Thallus** encrusting; forms irregular patches from 20–80mm across; blackish fruiting bodies occur towards centre.

**Distribution** Atlantic, English Channel and North Sea.

*Caloplaca marina*

### Anaptychia fusca (Hudson) Wainio

**Thallus** thick and bushy, growing in close-set lobes; black fruiting bodies.

**Distribution** Atlantic, English Channel and North Sea.

### Verrucaria maura Acharius

**Thallus** thin and encrusting, may cover extensive areas of rock ; resembles oil spills.

**Habitat** often associated with *Caloplaca* (see above).

**Distribution** Atlantic, English Channel and North Sea.

*Lecanora atra*

*Anaptychia fusca*

*Verrucaria maura*

## Marine angiosperms (sea-grasses)

The angiosperms, or flowering plants, are the most familiar representatives of the plant kingdom on land, and they include the grasses, herbs, shrubs and trees. Although a number of species are adapted to live partly or wholly in fresh water, very few can tolerate life in the sea. However, the Baltic with its low salinity provides a habitat for a number of plants that are associated with freshwater habitats, but these are really beyond the scope of this book. Those that have been included will generally be met with in fully maritime conditions, such as prevail on the coasts of the North Sea, the Atlantic and the Mediterranean, although some of these specialized marine angiosperms also occur in the Baltic. More details are available in Stace, C. 1997.

### *Zostera marina* Linnaeus
**Eel Grass, Sea-grass or Grasswrack**

(only a section of the leaf is illustated) Flat, long, narrow leaves rise for up to 1m, sometimes longer; leaf blade ends in a rounded tip: leaves between 5 and 15mm wide, veins shown in illustration; inconspicuous flowers, to some extent resembling those of terrestrial grasses, a perennial sea grass, most visible in Spring or Summer.

**Colour** dark green or grass green.

**Habitat** usually on sheltered beaches or estuaries, on gravel, sand and mud. Sometimes subtidal.

**Distribution** Mediterranean, Atlantic, English Channel, North Sea and Baltic. N.B. as the number of cross-references in this book will show, *Zostera* provides a natural habitat for other marine plants and animals.

section of
leaf of
*Zostera
marina*

### *Zostera angustifolia* (Horneman)
**Narrow-leaved Eel Grass**

(Not illustrated) regarded by some authorities as a variety of *Z. marina* (opposite). Similar to *Z. marina* with 3–5 veins and leaves 1–3mm wide. An annual plant found usually only between the tide marks especially in June–September.

*Zostera
marina*

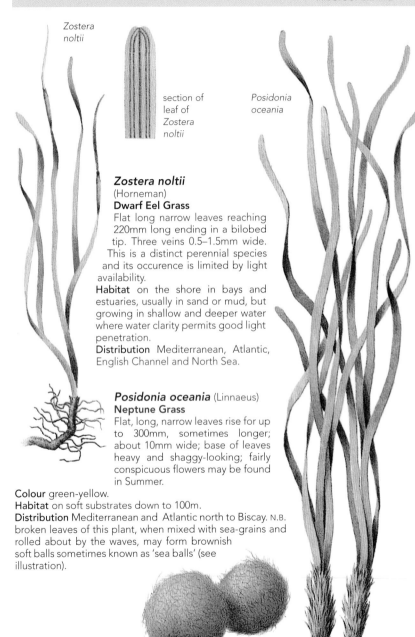

*Zostera noltii*

section of leaf of *Zostera noltii*

*Posidonia oceania*

### *Zostera noltii* (Horneman)
**Dwarf Eel Grass**
Flat long narrow leaves reaching 220mm long ending in a bilobed tip. Three veins 0.5–1.5mm wide. This is a distinct perennial species and its occurence is limited by light availability.
**Habitat** on the shore in bays and estuaries, usually in sand or mud, but growing in shallow and deeper water where water clarity permits good light penetration.
**Distribution** Mediterranean, Atlantic, English Channel and North Sea.

### *Posidonia oceania* (Linnaeus)
**Neptune Grass**
Flat, long, narrow leaves rise for up to 300mm, sometimes longer; about 10mm wide; base of leaves heavy and shaggy-looking; fairly conspicuous flowers may be found in Summer.
**Colour** green-yellow.
**Habitat** on soft substrates down to 100m.
**Distribution** Mediterranean and Atlantic north to Biscay. N.B. broken leaves of this plant, when mixed with sea-grains and rolled about by the waves, may form brownish soft balls sometimes known as 'sea balls' (see illustration).

Sea balls of *Posidonia*

65

# THE ANIMAL KINGDOM

▲ The seahorse *Hippocampus* being a fish is an example of a bilaterally symmetrical animal. Seahorse sightings around Britain have increased dramatically in recent years.

▼ The Sea Slater *Ligia* is another bilaterally symmetrical animal. It is a crustacean which moults and therefore is classified with the ecdysozoa.

The animal kingdom comprises an immense variety of organisms ranging from simple, one-celled creatures to complex, multicellular mammals. Between these extremes lies a broad spectrum of forms showing immense diversity.

Animals differ from plants in several fundamental ways. Firstly they are irritable and may respond to stimuli by movements rather than by growth. Secondly they are catabolic rather than anabolic, in other words, for their nutrition they rely on breaking down organic material into its constituent parts rather than elaborating organic compounds from inorganic ones. In this respect animals form themselves into very complex food chains which are founded upon the plants or primary producers, but which themselves are often interrelated; for example, carnivores preying upon herbivores. In any community there may be a chain or web of energy relationships, and this is as true in the sea as on land. In the marine environment, there is not the diversity of plant life that characterizes terrestrial habitats. The function that is fulfilled by the lowly algae is just as important, however. Whether they are the vast kelps of the shallow sea or the unicellular plants of the ocean plankton, they provide food for the herbivorous animals. On the seashore we find weed-eating sea-snails like the periwinkles; in shallow water there are the herbivorous sea-urchins and grazing fishes. Next in the chain are the primary carnivores such as whelks, starfishes, sea-anemones and fishes which in their turn feed on the herbivorous animals. In the ocean, minute crustaceans like copepods form the primary converters of plant protein into animal protein. In addition to finding a food supply, either as large macroscopic seaweeds or minute, suspended food particles, animals need shelter to protect them from their enemies and to provide them with the facilities for reproduction. In this respect other factors are as important as the supply of food in determining whether or not a particular species of animal can survive in a particular location. Different species are able to withstand different circumstances in terms of both the physical and the biological conditions for life. Thus it is that the seashore and shallow sea permit such a diversity of life; for the movements of the tides and the consequent alternation of exposure to air and sea provide an almost infinitely graduated range of environments from the terrestrial region immediately above the highest point to which the tides flow, down to the shallow seabed which is never exposed to the air.

The seashore is probably the finest training ground for the zoologist. The principles of ecology, as well as other aspects of interrelationships between and within species, can be appreciated in almost every type of habitat, and also provide examples of the greatest evolutionary interest. Although fewer species are recognized in the sea than in other environments, all types of animal known are represented there. Whilst some groups have conquered fresh water and land with moderate or great success, many have prospered only in the sea. Further-more some groups, such as the Ctenophora (comb-jellies or sea-gooseberries) and the Echinodermata (starfishes and sea-urchins), are exclusively marine.

The primary divisions or groups in the animal kingdom are known as *phyla* (singular phylum), and all the members of a phylum are considered by zoologists to be related to a common ancestral form by descent. The great majority of animals display bilateral symmetry (i.e. when divided in two from the head end towards the tail end they produce two halves which are mirror images of each other). When seeking to determine the evolutionary relationships of animals and thus to classify them, zoologists traditionally looked at their style of body architecture amd their pattern of embryonic development. Recent research on ribosomal RNA has shown that animals have affinities which do not entirely match the traditional view. Thus today we classify them according to shape (sponges are asymmetrical, sea anemones and their kin are radially symmetrical, whicle others, e.g. worms, crustacea and fish are bilaterally symmetrical). These bilaterally symmet-rical groups are then divided on the basis of their develop-mental embryology into deuterostomes (e.g. starfish and fish) and protostomes. Genetic work has revealed two major types of bilateral protostomes, those that moult (e.g. insects and crustacea, being called ecdysozoa) and those which have a particular larval type and feed with a particular style of tentacle crown, being called lophotrochozoa.

In the descriptions that follow, the text contains, as far as possible, details which complement the illustrations. A quick glance will show that in all cases the type of information that is provided is not exactly equivalent. Where this is the case it is largely a reflection of the state of the information currently available to scientists. In some cases it is known, for example, exactly how deep in the sea a particular species lives. In other cases this information is not known, and it is necessary to rely on a more general statement. It should also be remembered that the collector or beachcomber may find juveniles which are not as large as the adults generally encountered for the species.

Finally, as has already been pointed out, it is not possible in a book of this size to provide details of all the animals which might be encountered on the shore or in the shallow sea. However, as far as possible, references have been provided in the text and in the bibliography which should enable the reader to pursue the line of enquiry for a particular species a good deal further than is possible here.

▲ Radially symmetrical animals like this jellyfish *Rhizostoma* have no head or brain. They receive stimuli from all directions and have no front or back.

▼ The common octopus *Octopus* is a highly evolved bilaterally symmetrical animal with a brain and excellent sensory perception. It is one of the lophotrochozoa and is an efficient predator.

# PHYLUM PORIFERA SPONGES

These are sessile animals whose bodies consist of a single cavity with a major exhalent opening, and many smaller inhalent openings which are partly lined by special cells called choanocytes. The bodies also contain calcareous or siliceous spicules, or horny fibres, which provide support.

Sponges are the simplest animals treated in this book, yet for all their simplicity they are among the most difficult to describe carefully and scientifically. They are arranged in three classes: the Calcarea, with calcareous spicules; the Hexactinellida or Triaxonida, with siliceous spicules each of which has a six-rayed pattern; and the Demospongiae, with siliceous spicules which are never six-rayed, and/or horny fibres made of material known as spongin. It will be realized from these divisions that the nature and shape of the sponge spicules are of great importance when classifying and identifying sponges. Unfortunately these spicules cannot be seen without the aid of a microscope and the facilities for making the necessary microscopical preparations. However, it is possible to identify sponges to some degree by examining their gross form, texture and coloration, as well as by considering their habitat and distribution. Identifications based on these characters only can be provided here.

Sponges occur on almost all types of seabed from the lower shore down to the ocean depths. The form of their bodies depends to a great degree on the amount of exposure they have to face. Thus, in very exposed conditions sponges will be rounded or flattened against the substrate. In sheltered places they often assume a plant-like growth pattern with erect and delicate branching stems and shoots.

The basic body form of the sponge is illustrated on page 69. This diagram also indicates the movement of sea water through the sponge. Water supplies the animal with oxygen and food in the form of minute, suspended particles. This water enters the body by a number of pores and passes along the cavity or *paragaster* to leave by the large exhalent pore or *osculum*. Other sponges are more complex in their organization. The diagram on page 69 also shows how a compound arrangement may be formed from a number of units which become arranged round a communal paragaster and which share a central osculum. There are many variations on sponge architecture which are beyond the scope of this book, although most general text books on zoology provide good accounts. At first sight the paragaster may be compared with the gut of a higher animal, but it cannot strictly be regarded as such, for it lacks the specializations of tissues associated with a digestive tract. The cells which compose the sponge body are of very few types only, and those which line the paragaster absorb their own food requirements by directly ingesting suitable particles. Because of their level of organization, no organs can be identified within the sponge body which is thus regarded as being developed at the cellular grade. This lack of differentiation of cells into a number of tissue types making up organs means that sponges are capable of regenerating themselves from broken fragments. Despite

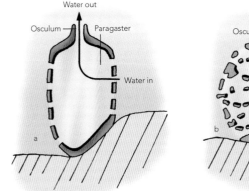 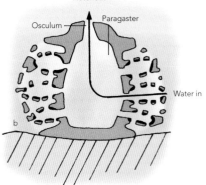

▲ Diagrammatic sections through (a) simple, and (b) complex sponges to show water currents and position of choanocytes (shown in heavy black).

their simple level of organization, sponges are able to undergo sexual reproduction with the formation of eggs and sperms which, after fertilization, give rise to a free-swimming larva. If the larva settles on a suitable substrate, it will develop into a new sponge.

An interesting feature of sponges is the extent to which they become associated with other organisms. If a large sponge is broken open, a number of other animals such as worms and crabs may be found residing in the cavity. Some sponges are associated with hermit crabs, giving camouflage and protection to the crab which in turn transports the sponge to new feeding areas.

Although a great number of sponge species are known, very few are of commercial importance. One exception is the Mediterranean species *Spongia officinalis,* the bath sponge, which is extensively collected from some of the Greek island waters.

The sponges are a notoriously difficult group to work on, and there are relatively few books available which help with their identification. The classification of calcareous sponges was brought up to date by Burton, M. 1963. *Revision of Classification of Calcareous Sponges.* British Museum (Natural History), London, Vosmeyer, G. C. J. 1935 *The Sponges of the Bay of Naples, Ponfera Incalcarea.* 3 vols. Martinus Niihoff, The Hague, describes the sponge fauna of Naples and the adjacent Mediterranean, but a number of the species descriptions are appropriate for the area covered by this guide. This work is beautifully illustrated.

## Class Calcarea

Sponges which contain calcareous spicules only. Their bodies are generally small and vase- or cup-shaped. They are generally pale in colour and grow away from light.

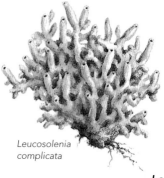

### *Leucosolenia complicata* (Montagu)

**Form** tubular, branching, up to about 20mm high; an osculum is found at the end of each branch; roughish texture but branches fragile.
**Colour** grey-white.
**Habitat** attached to seaweeds, hydroids, shells and ectoprocts from middle shore down to about 100m.
**Distribution** Atlantic, English Channel, North Sea and west Baltic.

*Leucosolenia complicata*

### *Leucosolenia botryoides* (Ellis & Solander)

**Form** similar to *L. complicata* (above), tubular but generally not branching, rising to about 20mm from a branching, root-like system of canals which adhere to the substrate; oscula is large.
**Colour** whitish.
**Habitat** attached to seaweeds and on rocks on the extreme lower shore and in shallow water.
**Distribution** Mediterranean, Atlantic, English Channel, North Sea and west Baltic.

*Leucosolenia botryoides*

### *Clathrina coriacea* (Montagu)

**Form** branching, encrusting, up to about 30mm high and 100mm wide as growths of twisted tubes; 1 osculum may suffice for several tubes where they join together.
**Colour** white-grey-yellow-brown-red.
**Habitat** under stones, rocks and shells, often in exposed places, from lower shore down to about 100m.
**Distribution** Atlantic, English Channel and North Sea.

*Clathrina coriacea*

### *Sycon ciliatum* (Fabricius)

**Form** tubular, erect, up to 30mm high, outer surface has a shaggy appearance under a hand lens; crown of longer, stiff spicules may be visible round the osculum at the apex.
**Colour** cream-yellow.
Habitat on rocks and shells from lower shore down to about 100m. *S. ciliatum* generally grows in clumps of several individuals, while *S. coronatum* is usually solitary. Some authorities believe there is only one species.
**Distribution** Mediterranean, Atlantic, English Channel and North Sea.

*Sycon ciliatum*

*Sycon ciliatum*
(mediterranean form)

### *Grantia compressa* (Fabricius)
**Purse Sponge**

**Form** growths may reach 50mm and are rounded and inflated under water; often collapses when taken out of water to form a flat, purse-like object with an opening which is the large terminal osculum.
**Colour** white-grey-yellow.
**Habitat** in groups, often under rocky overhangs and among red seaweeds on the lower shore and in shallow water.
**Distribution** Mediterranean, English Channel, Atlantic and North Sea.

*Grantia compressa*

71

# CLASS DEMOSPONGIAE

Sponges with siliceous spicules (never six-rayed) and/or horny fibres of spongin. Their bodies may be variously shaped, sometimes quite large. They are often brightly coloured and grow in well-illuminated places

### *Oscarella lobularis* (Schmidt)

**Form** encrusting, usually about 5mm thick and covering an area about 100mm wide; growth arranged in thick 'lobules' or lobes.

**Colour** pink-brown-yellow, occasionally blue-violet.

**Habitat** on stones, rocks and seaweeds from middle shore down to shallow water.

**Distribution** Mediterranean, Atlantic, English Channel and North Sea.

*Oscarella lobularis*

### *Pachymatisma johnstoni* (Bowerbank in Johnstone)
(Not illustrated)

**Form** grows to mound shaped masses, irregular, up to 150mm across. Smooth skin. conspicuous round exhalent openings arranged in irregular rows. Inside tissue is whitish.
**Colour** grey-blue.
**Habitat** on lower shore attached to rocks and in shallow water.
**Distribution** widely distributed in north west Atlantic area.

### *Suberites domuncula* (Olivi) *( =Ficulina ficus)*
**Sulphur Sponge** or **Sea-orange**

**Form** rounded, globular growth, up to 300mm in diameter but frequently less; texture fleshy; occasionally it is less round and encrusts rocks and piles, etc.

**Colour** orange-yellow.

**Habitat** from shallow water down to about 200m; rarely associated with whelk shells occupied by hermit crabs – the sponge may gradually dissolve away the whelk shell to provide the crab with shelter directly.

**Distribution** Mediterranean, Atlantic, English Channel and North Sea. N.B. avoid confusion with *Myxilla* (see page 75).

*Suberites domuncula*

### *Suberites camosus* (Johnstone)
(Not illustrated)

**Form** A smooth skinned round sponge, up to 150mm in diameter with a short stalk attaching it to rocks. One conspicuous terminal exhalent opening.
**Colour** orange-yellow-brown.
**Habitat** in shallow water on rocks.
**Distribution** Atlantic, English Channel and North Sea.

### *Cliona celata* Grant
**Boring Sponge**
**Form** smooth surfaced, branching, net-like.
**Colour** yellow-green-blue.
**Habitat** may occur on the surface of soft
rocks and shells as a marine rounded form
as well as boring into them so that they
become penetrated by a network of canals
filled with the sponge body: the outer surfaces of
the affected rocks and shells are easily recognized by the
many holes (each about 2mm in diameter) from which a
small portion of the sponge may project; also frequently
attacking dead *Ostrea* and *Venus* shells (see pages 173 and
181 ), from the lower shore down to about 100m.
**Distribution** Mediterranean, Atlantic, English Channel, North
Sea and west Baltic.

*Cliona
celata*

### *Tethya aurantia* (Pallas)
**Form** rounded, globular, up to 100mm in diameter;
outer surface covered with conspicuous warts;
osculum and pores about the same size. A cut
through the colony reveals a typical symmetrical
interior with an outer 'skin' and an inner 'core' a bit
like cutting through an orange.
**Colour** pale gold.
**Habitat** growing singly or in colonies on stones, rocks
and in caves from shallow water down to about 130m.
**Distribution** Mediterranean, Atlantic, English Channel and
North Sea.

*Tethya
aurantia*

### *Axinella verrucosa* (Esper)
**Form** erect, branching, up to about 250mm high;
form may be very variable.
**Colour** pink-gold-orange.
**Habitat** generally on hard substrates between 10
and 100m.
**Distribution** Mediterranean and Atlantic.

### *Axinella polypoides*
(Schmidt)
**Form** fan-like or erect and pillar-
like with a few branches reaching
up to 500mm high. Branches often
oval in cross section, oscula
have shallow surface grooves
radiating from them giving a
star-like effect; surface velvety
and smooth without ridges or
undulations.
**Colour** orange.
**Distribution** Atlantic.

*Axinella
verrucosa*

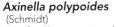

*Axinella
polypoides*

### *Hymeniacidon sanguinea* (Grant) *( =H. perleve)*

**Form** encrusting, with growths up to 500mm wide; many small oscula randomly placed; surface furrowed or smooth; form may be very variable.

**Colour** orange-scarlet-deep red.

**Habitat** on rocks from middle shore down to shallow water.

**Distribution** Atlantic, English Channel, North Sea and west Baltic.

*Hymeniacidon sanguinea*

### *Spongilla fluviatilis* (Pallas) *( =Ephydatia fluviatilis)*
#### Freshwater Sponge

**Form** branching, with irregular growths up to about 200mm; surface 'fuzzy' when seen in detail.

**Colour** green; small, rounded, brown-orange overwintering bodies may be visible in the tissue at certain times of the year.

**Habitat** attached to tree roots and stones, etc., in fresh water.

**Distribution** in many European rivers and in the Baltic Sea where the salinity is sufficiently low to allow it to grow.

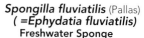

*Spongilla fluviatilis*

### *Halichondria panicea* (Pallas)
#### Breadcrumb Sponge

**Form** often growing in large encrustations up to 200mm across and about 20mm thick; the many oscular openings resemble craters set in minute volcanic cones; surface texture fairly smooth; form may be variable.

*Halichondria panicea*

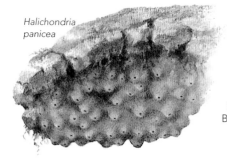

**Colour** white-orange-yellow-green-brown, occasionally red.

**Habitat** on stones, seaweeds and shells from middle shore down to quite deep water.

**Distribution** Mediterranean, Atlantic, English Channel, North Sea and west Baltic.

### *Myxilla incrustans* (Esper)
**Form** encrusting, forming thick, cushion-like growths up to 150mm across and 50mm thick; the large oscula are arranged unevenly over the furrowed surface.
**Colour** yellow-brown-orange.
**Habitat** on rocks, or sometimes on hydroids or spider crabs from lower shore down to about 130m.
**Distribution** Mediterranean, Atlantic, English Channel and North Sea. N.B. avoid confusion with *Suberites domuncula* (see page 72).

*Myxilla
incrustans*

### *Verongia aerophoba* (Schmidt)
### ( *=Aplysina aerophoba)*
**Form** erect, pillar-like and cylindrical, up to 150mm high; the flattened tips, in which lie the terminal oscula, have the appearance of being 'sawn off'; individuals join at base; the skeleton lacks spicules and consists of spongin fibres; hard nodules may be embedded in the tissue.
**Colour** yellow-green-black.
**Habitat** generally on rocks in shallow water.
**Distribution** Mediterranean, and Atlantic north to Biscay.

*Verongia
aerophoba*

### *Dysidia fragilis* Montagu
**Form** up to 40mm high, rounded and cyclindrical with large conspicuous oscula grouped terminally; sometimes encrusting; several may grow side to side. Surface is covered in small pointed outgrowths; texture variable, often soft and elastic; ostia are small and flush with the surface.
**Colour** whitish.
**Habitat** attached to rocks and often growing in crevices.
**Distribution** Mediterranean, Atlantic west to English Channel and Ireland.

*Dysidia
fragilis*

# PHYLUM CNIDARIA

This phylum is composed of generally soft-bodied, flower-like animals, often with a jellyfish stage in the life cycle.

The Cnidaria includes some of the most beautiful and abundant marine animals. They are of simple form, having sac-like bodies composed of an inner layer of cells (endoderm) surrounded by an outer layer (ectoderm). These two layers are separated by a jelly-like layer called the mesoglea. The interior of the sac acts as a stomach and opens to the outside via the mouth which also serves for an anus. Around the mouth are rings of tentacles armed with stinging cells. These tentacles grasp the prey and, after it has been immobilized by the stinging cells, force it into the mouth. There are no circulatory nor excretory systems, and the nervous system is very simple. Each animal is termed a *polyp,* unless it is a jellyfish when it is usually called a *medusa.*

Three classes are recognized. Members of the class Hydrozoa (sea-firs) are hydroids which occur in most habitats from the shore to the deep sea, and they are the simplest cnidarians. The polyps often grow close together and are interconnected by tubular extensions of the stomach (see diagram on page 77). This arrangement frequently leads to the formation of colonies in which some of the polyps are specialized for feeding, whilst others are responsible for reproduction or defence. The form and growth pattern of the colony are important identification guides. In some types only the tubular inter-gastric connections are surrounded by a protective skeletal sleeve (the perisarc) – these are the order of *athecate* hydroids. Elsewhere the perisarc extends up around the gastric region and provides a protective cup (the theca) which houses the whole polyp and into which the tentacles can be withdrawn, Such hydroids are known as *thecate,* and are placed in another order. Hydroid life cycles are complicated. Adult colonies can grow on stones, shells or seaweeds. Some may develop non-feeding, reproductive polyps from which free-swimming medusae bud off. These are swept away by tides and currents and can only swim upwards and float downwards. The umbrella-shaped bell (see diagram below) contracts rhythmically, lifting the medusa in the water. Sex organs are developed by the medusae and when the sperms and eggs are ripe, they are released into the sea where fertilization occurs. A larva is formed which eventually settles on the seabed to form a new colony. Early naturalists did not understand this life cycle and so they gave the hydroids different names from their respective medusae. Both hydroids and medusae generally feed on small organisms which collide with their sting-celled tentacles. A third order in the class Hydrozoa is the Siphonophora. These are large, floating, colonial hydrozoans which differ from other members of the class in that they are free living for the whole of their life cycle. One specialized polyp forms the *float* and is surrounded by many others developed for feeding, reproduction or defence. Many siphonophores are surface animals relying on wind and water currents for movement. They catch small or larger prey with their long, trailing tentacles, and usually appear in European waters when strong south-

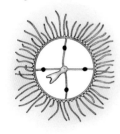

▼ Free swimming medusa of *Obelia geniculata*

westerly winds sweep them in from the Atlantic. Their long tentacles may be broken off when they are cast up on the shore. The small order Chondrophora consists of a few species of free-floating hydroid polyps which produce medusae.

In the class Scyphozoa (jellyfishes) the medusa stage dominates, and individuals spent most of their lives as floating predators catching their prey with their long, trailing tentacles equipped with stinging cells. The ripe gonads of the jellyfish release sperms and eggs and, after fertilization, a small larva forms which settles on the seabed to develop a small, hydroid-like organism (the *scyphistoma*). This stage buds repeatedly, giving rise to miniature jellyfishes which then develop into the characteristic adults. In one group (stalked jellyfishes) the small, trumpet-shaped adults are not free swimming, but live attached to the fronds of seaweeds and stones. Scyphozoans occur at most depths in the shallow seas.

The class Anthozoa (sea-anemones and their allies) is a diverse group with no medusa stage. The polyps are more sophisticated than the hydrozoan type and can either burrow in soft substrates or live attached to rocks and shells. The best-known anthozoans are the sea-anemones which belong to the order Actinaria. Examples are to be found on the shore as well as in deeper water. Like other cnidarians they are carnivorous, catching their food with tentacles. They can reproduce asexually by division, or sexually when a fertilized egg develops into another polyp – usually via a larval stage. The order Antipatharia comprises the black corals. These have colonial polyps spread out over a hard, horny, tree-like skeleton. The Ceriantharia and Zoantharia resemble sea-anemones. The former are rather worm-like, and live in slimy tubes buried in mud or sand from which their long tentacles protrude, the latter are small and colonial, and are usually found encrusting rocks and stones. The true corals (order Madreporaria) are represented by a few species in the European seas. They differ from anemones because they have hard, chalky skeletons which support and protect the lower regions of the polyp, and into which the tentacles can usually be drawn. Many are colonial, unlike the European representatives of the closely related Corallimorpharia which, although they resemble coral polyps in many ways, lack the characteristic skeleton of limestone. The orders Alcyonacea, Gorgonacea and Pennatulacea are closely related and usually have polyps with branching tentacles. They are always colonial and have a variety of growth forms.

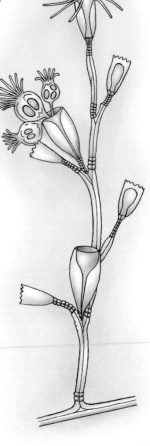

▼ Branch of *Gonothyraea loveni* to show the form of feeding and reproductive polyps. In this species medusae are not liberated, but remain attached to the colony.

## Class Hydrozoa

### Order Athecata

Hydroids in which the horny perisarc does not surround the polyps. The polyps can always be seen even when disturbed. For identification of other hydroids see Hincks, T. 1868. Russell, F. S. 1953 and 1970 gives a detailed account of hydromedusae.

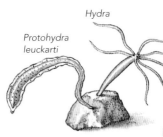

*Hydra*

*Protohydra leuckarti*

#### *Protohydra leuckarti* Greef

Solitary; up to 2mm high; lacking perisarc and tentacles.

**Habitat** brackish water; often attached to dead seaweeds in muddy areas.

**Distribution** Baltic.

#### *Hydra* species

Solitary; up to 20mm high; possessing tentacles but no perisarc.

**Colour** brown-white-green according to species.

**Habitat** on water plants in sheltered places; characteristic of fresh water.

**Distribution** Baltic where salinity is low.

*Tubularia indivisa*

#### *Tubularia indivisa* Linnaeus

Colonial with creeping 'roots' and erect, seldom branching stems often tightly plaited; up to 180mm high; terminal, flask-shaped, red-pink polyps bear an outer ring of drooping white tentacles and a shorter, stiffer, inner ring; occasionally with reproductive bodies resembling miniature bunches of grapes; perisarc often striped longitudinally and yellowish.

polyp

**Habitat** in rock pools on the lower shore and attached to rocks and boat wrecks in deeper water.

**Distribution** Atlantic, English Channel and North Sea. N.B. a number of closely related species occur in these areas and in the Mediterranean.

#### *Tubularia larynx* Ellis and Solander

Colonial, stems less than 50mm high and branching, with outer perisarc ornamented by a series of rings after each branch, many short tentacles near mouth of feeding polyp, many longer tentacles away from mouth.

**Habitat** as for *T. indivisa* (above).

**Distribution** Atlantic, English Channel and the North Sea.

#### *Coryne pusilla* Gaertner

Colonial with creeping root-like stolons and irregularly branching stems, up to 100mm high. Terminal cylinder-shaped pink polyps bear club-

*Tubularia larynx*

shaped tentacles; occasionally found with spherical reproductive bodies.
**Habitat** on rocks and in pools, lower shore.
**Distribution** Atlantic, English Channel and North Sea.

### Sarsia eximia (Allman)

Colonial, a creeping root-like stolon gives rise to upright stems reaching 100mm high. The perisarc is marked by a series of rings after each junction. Polyps are narrow and quite long bearing 30 or so short club-tipped tentacles. Reproductive polyps born at base of each tentacle.
**Habitat** on algae or rocks on the lower shore and in shallow water, sometimes on buoys and floats.
**Distribution** Atlantic, English Channel and North Sea.

### Cladonema radiation Dujardin

Colonial with creeping root-like stolons from which arise unbranching stalks each bearing a small polyp; the polp has four short basal tentacles and four large club-shaped tentacles near the mouth; reproductive bodies may arise on the sides of the polyp near the lower tentacles; polyp about 5mm high; perisarc inconspicuous.
**Habitat** on rocks and pebbles on the shore and in shallow water.
**Distribution** Atlantic, English Channel and North Sea. N.B. often appears in marine aquaria where its relatively large medusa (3.5mm) may develop and be seen swimming or adhering to the glass.

### Clava multicornis (Forskål)

Colonial with an open network of 'roots'; short, unbranching stems rising to 13mm; terminal, elongated, pink polyps with tapering tentacles; occasionally found with spherical reproductive bodies.
**Habitat** in rock pools on lower shore and on stones and seaweeds in shallow water.
**Distribution** Atlantic, English Channel, North Sea and Baltic.

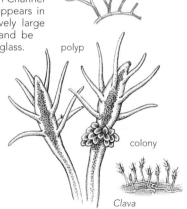

*Coryne pusilla* — colony

*Coryne pusilla*

*Sarsia eximia* — polyp

*Cladonema radiation*

polyp

colony

*Clava multicornis*

79

### Hydractinia echinata (Flemming)

Colonial; densely growing with an encrusting perisarc; spindle-shaped, white-brown-red polyps rising on stems up to 15mm high.

**Habitat** on *Buccinum* shells (see page 161) inhabited by hermit crabs, and sometimes on stones in shallow water.

**Distribution** Mediterranean, Atlantic, English Channel, North Sea and occasionally in the Baltic.

*Hydractinia echinata*

colony

### Eudendrium rameum (Pallas)

Colonial and bush-like; up to 150mm high with creeping roots; strong stem with brown perisarc; flask-shaped pink polyps.

**Habitat** in water usually deeper than 10m, on rocks and in caves.

**Distribution** Mediterranean, Atlantic, English Channel and rarely in the North Sea and Baltic.

### Bougainvillia ramosa
(Van Beneden)

**Medusa generation** rounded and bell-shaped; up to 4mm high; 4 groups of 4–9 tentacles; mouth stalk and gonads (when ripe) green-brown.

**Habitat** pelagic.

polyp

**Hydroid generation** colony reaching up to about 50mm high, spindle-shaped polyps with up to 12 pale tentacles; perisarc yellow-brown.

**Habitat** on stones and other hydroids from lower shore down to 30m.

**Distribution** Mediterranean, Atlantic, English Channel, North Sea and west Baltic.

*Eudendrium rameum*

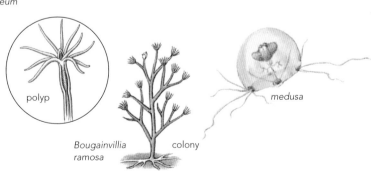

polyp

*Bougainvillia ramosa*   colony   medusa

### Rathkea octopunctata (M. Sars)
Medusa of the hydroid **R. octopunctata** (Hydroid not illustrated). Blunt, cone-shaped umbrella reaching about 5mm high; generally with 8 clusters of 4 tentacles arranged round the bottom of the umbrella.
**Habitat** pelagic.
**Distribution** Atlantic, English Channel, North Sea and Baltic.

*Rathkea octopunctata*

## ORDER THECATA

Hydroids in which the horny perisarc extends around the polyps, so providing a cup into which they can be withdrawn from view when disturbed. Most species have a characteristic growth pattern (avoid confusion with some of the Ectoprocta, see pages 232–237). For identification of other hydroids see Cornelius, P. F. S. 1995 and Hincks, T. 1868. Russell, F. S. 1953 and 1970 gives a detailed account of hydromedusae.

### Obelia geniculata (Linnaeus)
Colonial with creeping 'roots' and erect, zig-zag, branching stems reaching 40mm high; bell-shaped polyp cups supported on a red-brown perisarc which is characteristically ringed at branch points; polyps borne on alternate sides of stem.
**Habitat** on seaweeds on the lower middle shore and lower shore, and in shallow water.
**Distribution** Mediterranean, Atlantic, English Channel, North Sea and Baltic.

colony

detail of colony

*Obelia geniculata*

### Dynamena pumila (Linnaeus)
Colonial, stems arising from horizontal creeping root-like stolons; up to 12mm tall, polyps born on opposite sides of upright stems, polyp cups are cowl-shaped.
**Habitat** on rocks and weed on the shore and in shallow water, often very common.
**Distribution** Atlantic, English Channel and North Sea.

Polyp

*Dynamena pumila*

*Gonothyraea loveni*

detail of colony

### Gonothyraea loveni (Allman)
Colonial with creeping 'roots' and erect stems which reach up to 30mm high; polyp cups narrower than in *Obelia* (above) and fringed with small teeth.
**Habitat** on seaweeds, e.g. *Fucus* (see pages 38–40), on the lower middle shore and lower shore, and in shallow water.
**Distribution** Mediterranean, Atlantic, English Channel, North Sea and Baltic.

colony

### Aequorea aequorea (Forskål)

Probably the medusa of *A. paracuminata* (Not illustrated). Cap-shaped medusa reaching up to 100mm in diameter, occasionally much larger; up to 400 long, pale tentacles; mouth with red lips; gonads blue-pink when mature.
**Habitat** pelagic.
**Distribution** Mediterranean, Atlantic and North Sea.

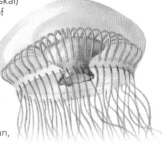

*Aequorea aequorea*

### Sertularia cupressina (Linnaeus)
#### White Weed

Colonial with creeping 'roots' and relatively large branching colonies reaching up to 450mm high; main stem often twisted and frequently branched so that polyps occur in tufts arranged in 2 rows with longish, cylindrical cups; colonies white or pink.
**Habitat** on stones, shells (e.g. *Aequipecten opercularis*, see page 180) and sometimes on crabs, in shallow and deeper water.
**Distribution** Atlantic. English Channel, North Sea and west Baltic

detail of colony

*Sertularia cupressina*

### Plumularia catharina Johnston

Colonial with creeping 'roots' and branching stems which reach up to 10 high; clear identifying character is the arrangement of the side-branches which are exactly opposite; polyp cups are quite deep with an even, untoothed margin and are borne on the main stem as well as on the side-branches.
**Habitat** on stones, shells and the tunics of sea squirts (see pages 272–275).
**Distribution** Atlantic, English Channel and North Sea.

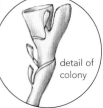

detail of colony

colony

### Plumularia setacea (Ellis and Solander)

Colonial; delicate wavy stems with creeping 'roots' arise from creeping stolons, reaching about 30mm high; polyps are born on side branches which are alternatively arranged; polyp cups quite deep and smooth-lipped. Slender pod-like reproductive bodies may be born in the angle of the joint of the side branch and main stem.

*Plumularia catharina*

Habitat lower shore and shallow water often growing on other organisms.
Distribution Atlantic, English Channel and North Sea.

*Plumularia setacea*

### *Campanularia hincksi* Alder

Colonial with conspicuous, elongated polyp cups which are castellated round the margin and supported on a long branch; up to 5mm high from the base stem which creeps over the substrate; long, elongated reproductive polyps may arise directly from the base stem, being ringed and tapering towards the opening at the top (not illustrated here).

detail of colony

*Campanularia hincksi*

Habitat on shells, other hydroids and ectoprocts from about 10–60m deep.
Distribution Mediterranean, Atlantic, English Channel and North Sea.

*Nemertesia antennina*

### *Nemertesia antennina* (Linnaeus)
### ( =*Antennularia antennina*)

Colonial, clusters of upright stems reaching 250mm high; yellowish horny texture. The stems arise from a matted filamentous wall which serves as a holdfast; the whorled short branchlets are wide at the base, curved in, bearing polyps which are housed in vase-like polyp cups with a smooth rim; reproductive bodies occur in the angle of the main stem and branchlet.

detail of colony

Habitat usually attached to rigid objects e.g. pebbles and empty shells although often where sand accumulates in shallow and deeper water.
Distribution Atlantic, English Channel and North Sea.

### *Nemertesia ramosa* (Lamouroux)

Colonial, the thick main stem arises from a holdfast of matted fibres and divides and sub-divides irregularly; the long tapering outcurved branchlets are hairy and closely set and arranged in whorls; the polyp cups are small and vase-like; the reproductive bodies are pear shaped and face towards the stem.
Habitat and Distribution as for *N. antennina*.

detail of colony

*Nemertesia ramosa*

83

## ORDER SIPHONOPHORA

Colonial, free-floating or swimming hydrozoans with many individuals variously modified to provide floats, stems linking the polyps, and feeding, defensive and reproductive individuals. Some species float on the surface, others swim in deeper water. They are generally oceanic but are occasionally washed into coastal areas by rough weather. Kirkpatrick, P. A. and Pugh, P. R. 1984 give more details of this group.

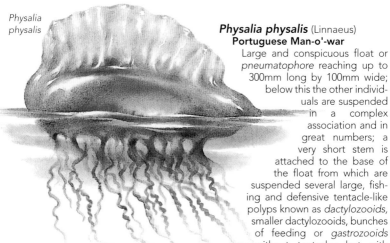

*Physalia physalis*

### *Physalia physalis* (Linnaeus)
**Portuguese Man-o'-war**

Large and conspicuous float or *pneumatophore* reaching up to 300mm long by 100mm wide; below this the other individuals are suspended in a complex association and in great numbers; a very short stem is attached to the base of the float from which are suspended several large, fishing and defensive tentacle-like polyps known as *dactylozooids*, smaller dactylozooids, bunches of feeding or *gastrozooids* without tentacles but with mouths, and many branched reproductive individuals or *gonodendra*; gonodendra may release medusae which form part of the life cycle like the medusae of the other hydroids.

**Colour** pneumatophore silver blue with red tinging; rest of colony blue-purple.

**Habitat** pelagic, surface dweller.

**Distribution** Mediterranean and Atlantic. N.B. when rarely cast ashore the pneumatophore may be broken from the remaining part of the colony which thus appears missing. Beware dangerous stinging cells.

*Muggiaea atlantica*

### *Muggiaea atlantica* Cunningham

Helmet-shaped swimming bell up to 20mm long; transparent; retractile stem hangs from the bell and supports a great number of combinations of individuals; each group is known as a *cormidium* and consists of a feeding zooid equipped with 1 tentacle, a reproductive zooid and a small swimming bract (like a medusa) which enables the reproductive zooids to lead a free existence.

**Habitat** found swimming at various depths.

**Distribution** Mediterranean and Atlantic.

### Lensia conoidea (Keferstein & Ehlers)
(Not illustrated) Similar to *Muggiaea atlantica* (opposite) but equipped with 2 swimming bells each about 10mm long; front bell is slightly larger than the rear; retractile stem bears numerous cormidia.
**Habitat** found swimming at various depths, often in shoals.
**Distribution** Mediterranean, Atlantic and North Sea.

*Physophora hydrostatica*

### Physophora hydrostatica Forskål
Small, apical float or pneumatophore below which hangs a stem polyp about 60mm in length supporting 2 rows of swimming bells; below these trail feeding, defensive, fishing and reproductive polyps.
**Colour** predominantly yellow-pink-red.
**Habitat** found swimming at various depths.
**Distribution** Mediterranean, Atlantic, very occasionally English Channel and North Sea.

## ORDER CHONDROPHORA
Free-floating, colonial hydrozoans of uncertain position in the hydrozoa, regarded by name as colonial and by others as solitary.

### Velella velella (Linnaeus)
**By-the-wind-sailor**
Modified siphonophore with a bluish, round or oval disc reaching 80mm in diameter, which encloses the float and contains a horny skeleton equipped with a sail; when alive the sail is covered with soft tissue and projects above the surface of the water to catch the wind and aid dispersal; large feeding zooid under the disc is encircled by a ring of reproductive zooids; at the periphery is a larger ring of tentacle-like fishing zooids.
**Habitat** pelagic; surface dweller; sometimes in shoals.
**Distribution** Mediterranean and Atlantic.

*Velella velella*

### Porpita umbella Otto
(Not illustrated) Similar to *Velella velella* (above) but without the characteristic sail; blue-green disc reaching up to 80mm in diameter.
**Habitat** pelagic, surface dweller, often in shoals.
**Distribution** Mediterranean and Atlantic.

# Class Scyphozoa Jellyfishes

Cnidarians in which the medusa stage is dominant, but which normally pass through a small polyp phase during the life cycle; this polyp is termed the *scyphistoma*. The medusae are mostly pelagic, but a few are sessile. Russell, F. S. 1970 gives a detailed account of scyphozoa.

*Haliclystus auricula*

### *Haliclystus auricula* (Rathke)
#### Stalked Jellyfish or Sessile Jellyfish

Trumpet-shaped body; up to 50mm high; 'bell' drawn out into 8 lobes each bearing tentacles; mouth with 4 corners; conspicuous, wart-like 'anchor' arranged between each lobe (these are used for temporary attachment).

**Habitat** attached by means of adhesive stalk to seaweeds or sea-grasses in rock pools on lower shore and in shallow water.
**Distribution** Atlantic, English Channel and, rarely, North Sea and west Baltic. N.B. a closely related species *Lucernariopsis campanulata* may be found; this lacks the wart-like anchors. Four other similar species may be found.

### *Pelagia noctiluca* (Forskål)

'Umbrella' mushroom-shaped; transparent, but tinted yellow-red; up to 100mm in diameter; 4 arms around mouth; 8 slender trailing tentacles around periphery; tentacles longer than mouth arms when fully extended; 8 small, wart-like sense organs alternate with tentacles.
**Habitat** pelagic.
**Distribution** Mediterranean, Atlantic and English Channel. N.B. can inflict severe and painful stings. Luminescent when disturbed at night.

*Pelagia noctiluca*

*Chrysaora hysoscella*

### *Chrysaora hysoscella* (Linnaeus)
#### Compass Jellyfish

'Umbrella' saucer-shaped; up to 300mm in diameter and drawn out into 32 lobes at periphery, and bearing 24 tentacles alternating with 8 sense organs; 4 mouth arms longer than tentacles.
**Habitat** pelagic.
**Distribution** Atlantic, English Channel and North Sea.

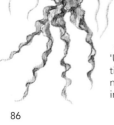

### *Cyanea lamarckii* Péron & Lesueur

'Umbrella' saucer-shaped; up to 150mm in diameter, sometimes more; drawn out into 32 lobes at periphery and bearing numerous tentacles (not all illustrated for simplicity) arranged in 8 clusters; 4 frilly mouth arms shorter than tentacles.

Colour blue-white.
Habitat pelagic.
Distribution Atlantic north from Biscay, English Channel and North Sea. N.B. can inflict severe stings.

### *Cyanea capillata* (Linnaeus)
**Lion's Mane Jellyfish**
(Not illustrated) Similar to *C. lamarckii* (above). Up to 500mm in diameter, but occasionally more.
Colour brick-red-yellow.
Habitat pelagic.
Distribution Atlantic north from Biscay, English Channel, North Sea and west Baltic.

*Cyanea
lamarckii*

### *Aurelia aurita* (Linnaeus)
**Common Jellyfish**
'Umbrella' saucer-shaped; up to 250mm in diameter; frilly mouth arms longer than the numerous short tentacles; 8 sense organs; 4 conspicuous purple-violet reproductive organs which are horseshoe-shaped when seen from above.
Colour transparent; tinged blue-white.
Habitat pelagic.
Distribution Mediterranean, Atlantic, English Channel, North Sea and Baltic. N.B. inset illustration shows *scyphistoma* (see right) which is found attached to rocks and seaweeds in pools and shallow water; from this the small larvae with bilobed arms bud off; they are known as *ephyrae*.

*Aurelia
aurita*

### *Rhizostoma pulmo* (Macri) *( =R. octopus)*
'Umbrella' domeshaped; up to 900mm in diameter; no peripheral tentacles; 96 edge lobes and 16 sense organs; 8 fused mouth arms.
Colour  blue-white-yellow, with yellow or blue-red mouth arms.
Habitat pelagic.
Distribution Mediterranean, Atlantic, North Sea and west Baltic.

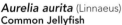

*Rhizostoma
pulmo*

87

## CLASS ANTHOZOA Sea-anemones and their allies

Cnidarians which lack a medusa stage in their life cycle. The polyps may be solitary or colonial, and are often large and conspicuous; a chalky skeleton may or may not be present. See Manuel, R. L. 1988 for a detailed account of this group.

### ORDER ANTIPATHARIA *BLACK CORALS*

The thorny, black branching is a skeleton surrounded by softer tissues bearing polyps which cannot retract their tentacles, the tentacles are unbranched.

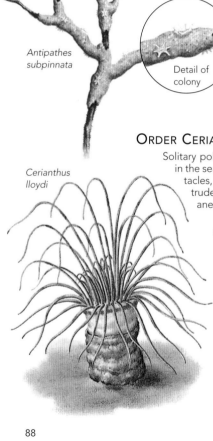

*Antipathes subpinnata*

Detail of colony

#### *Antipathes subpinnata* (Ellis & Solander)
**Black Coral**

Colonial, with a black-brown skeleton up to 1m high; white-grey outer tissue with small, bilaterally symmetrical polyps up to 1mm high.

**Habitat** on muddy substrates with stones between 10 and 250m.

**Distribution** Mediterranean and Atlantic as far north as English Channel approaches.

### ORDER CERIANTHARIA

Solitary polyps living in thick, mucuous tubes buried in the seabed so that the crown of unbranched tentacles, which are arranged in 2 whorls, can protrude. There is no adhesive basal disc as in true anemones.

*Cerianthus lloydi*

#### *Cerianthus lloydi* Gosse

Polyp up to 200mm high, often less; long, yellow body bearing about 60 brown, peripheral tentacles reaching up to 40mm; similar number of shorter, inner tentacles; animal characteristically withdraws into its tube with a rapid jerk if disturbed; tube slimy at top, more felt-like at base.

**Habitat** burrowing in muddy sand from 1–35m.

**Distribution** Atlantic, English Channel and North Sea.

### Pachycerianthus multiplicatus
Carlgren
**Fireworks Anemone**
A large polyp reaching up to 300mm high
and with a tentacle span of 300mm. Inner
short tentacles surround mouth and are
light brown in colour. Outer tentacles are
long and banded brown and white. Tube is
very long up to 1m.
**Habitat** tube embedded in mud or muddy
sand from around 10m down.
**Distribution** Atlantic north to Scandinavia.

*Pachycerianthus
multiplicatus*

## ORDER ZOANTHARIA
### ENCRUSTING 'ANEMONES'

These polyps are usually colonial, and are linked by a stolon
which may or may not be easily seen. They encrust existing
surfaces such as rocks and shells, and have smooth, slender
tentacles which sometimes appear to have minute terminal
expansions.

### Epizoanthus couchi (Johnston)
Colonial, with the stolons rather obscure;
grey-brown polyps up to 10mm high; about
35 slender tentacles reaching 5mm long.
**Habitat** on rocks, stones and pebbles from
extreme lower shore down to deep water.
**Distribution** Atlantic, English Channel and
North Sea.

*Epizoanthus
couchi*

### Epizoanthus papillosus (Johnstone)
(Not illustrated) Similar to *E. couchi* (above), but usually found
on shells inhabited by the hermit crabs from 10–100m.
**Distribution** Atlantic, English Channel and North Sea.

### Parazoanthus axinellae (O. Schmidt)
Colonial, with stolons as thin lamellae con-
necting bright yellow polyps which reach
10mm high; about 30 tentacles.
**Habitat** on stones and other organisms,
e.g. sponges, from 6m downward.
**Distribution** Atlantic, English Channel and
North Sea.

### Parazoanthus anguicomus (Norman)
(Not illustrated) Similar to *P. axinellae* (above) but
slightly larger with whitish polyps up to 10mm high.
**Habitat** on cave walls, rock faces, sponges and sea-squirts
from 6m downward.
**Distribution** Atlantic north of UK and Scandinavia.

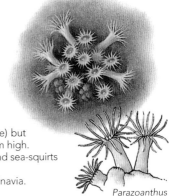

*Parazoanthus
axinellae*

## ORDER ACTINARIA *SEA-ANEMONES*

Solitary, often conspicuous anthozoans with no calcareous, hard skeleton. They are sedentary, with the polyp base modified either for burrowing, or for adhering to rocks and shells by means of an adhesive, sucker-like disc. The tentacles are simple and unbranched. For identification of other anemones see Stephenson, T. A. 1928 and 1935, or Manuel, R. L. 1988.

### *Edwardsiella carnea* (Gosse) ( *=Milne-Edwardsia carnea)*

*Edwardsiella carnea*

No adhesive basal disc; not more than 20mm high when extended; worm-like translucent pink body; about 30 delicate tentacles.
**Habitat** occupies small holes (made by other organisms) in rocks, overhangs, cave walls, etc. on lower shore and in shallow water.
**Distribution** Atlantic, English Channel and North Sea; usually very local.

whole animal

*Edwardsia claparedii*

### *Edwardsia claparedii* (Panceri)

No adhesive basal disc; not more than 100mm high when fully extended; worm-like, translucent pink body; 16 delicate tentacles, transparent and marked with fine dots; diameter across tentacles when expanded 35mm
**Habitat** burrowing in muddy sand and gravels (often around *Zostera* beds, see pages 64–65) on extreme lower shore and down to 10m.
**Distribution** Mediterranean, Atlantic, English Channel and North Sea

### *Halcampa chrysanthellum* (Peach)

No adhesive basal disc; not more than 50mm when fully extended; worm-like, translucent buff-coloured body, tinged yellow above; 12 short, stout tentacles with a red-brown horseshoe mark at the base.
**Habitat** in sandbanks and among *Zostera* beds (see pages 64–65), sometimes under stones or in pools from lower shore down to 100m.
**Distribution** Atlantic, English Channel, North Sea and Baltic.

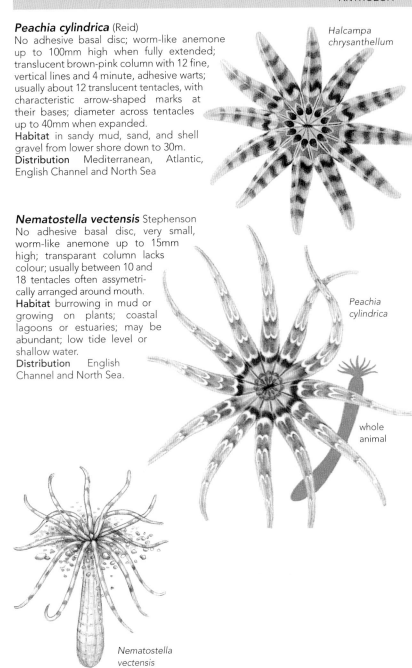

### *Peachia cylindrica* (Reid)

No adhesive basal disc; worm-like anemone up to 100mm high when fully extended; translucent brown-pink column with 12 fine, vertical lines and 4 minute, adhesive warts; usually about 12 translucent tentacles, with characteristic arrow-shaped marks at their bases; diameter across tentacles up to 40mm when expanded.

**Habitat** in sandy mud, sand, and shell gravel from lower shore down to 30m.

**Distribution** Mediterranean, Atlantic, English Channel and North Sea

### *Nematostella vectensis* Stephenson

No adhesive basal disc, very small, worm-like anemone up to 15mm high; transparent column lacks colour; usually between 10 and 18 tentacles often assymetrically arranged around mouth.

**Habitat** burrowing in mud or growing on plants; coastal lagoons or estuaries; may be abundant; low tide level or shallow water.

**Distribution** English Channel and North Sea.

*Halcampa chrysanthellum*

*Peachia cylindrica*

whole animal

*Nematostella vectensis*

91

### *Actinia equina* Linnaeus
**Beadlet Anemone**

red form

Base adhesive and sucker-like; smooth column up to 70mm high and 60mm across when fully expanded; about 200 densely packed retractile tentacles which can reach 20mm in length and are arranged in 5–6 circlets; they fold in quickly when the animal is disturbed; 24 relatively conspicuous blue spots arranged on the periphery of the oral disc outside the tentacles; when the tide is out it may appear as a blob of jelly up to 30mm high.

*Actinia equina*

green form

**Colour** variable; brown, red, orange or green.

**Habitat** common on rocks and in crevices from the middle shore down to 8m.

**Distribution** Mediterranean, Atlantic, English Channel and North Sea.

*Actinia fragacea*

### *Actinia fragacea* Tugwell

Similar to *A. equina* (above) but larger reaching 100mm in diameter.

**Colour** Column red with greenish yellow spots.

**Habitat** lower shore among rocks and in shallow water.

**Distribution** Atlantic and English Channel.

### *Anemonia viridis* (Forskål)
**Opelet**

Base lightly adhesive and sucker-like; smooth column reaching up to 100mm high, but often squat with base spread

*Anemonia viridis*

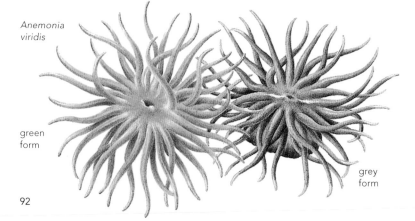

green form

grey form

out; about 170 wavy tentacles which may reach 150mm in length, and arranged in up to 6 circlets; tentacles cannot be fully retracted.

**Colour** column brown, grey or green; tentacles similar with or without purple tips; characteristically 2 white lines run from opposite sides of the disc to the mouth.

**Habitat** on the lower shore and down to 23m on rocks and, occasionally, seaweeds; generally prefers strong light.

**Distribution** Mediterranean, Atlantic north to west Scotland, and English Channel east to the Solent.

### *Urticina felina* (Linnaeus)

Base strongly adhesive and sucker-like; warty column reaching up to 100mm when large specimens are fully expanded; characteristically with pieces of shells and gravel adhering to the column so that when closed the animal may be quite inconspicuous; warts conspicuous on column; base up to 55mm in diameter; 80–160 stout, retractile tentacles with a span when extended of 200mm.

*Urticina felina*

**Colour** body grey, blue or green with irregular red bands or patches; mouth and oral disc olive-green, pink or blue; coloration very variable and some combinations may be outstandingly beautiful; tentacles translucent and banded with white, grey, green, blue or red shades.

**Habitat** on hard substrates often in crevices and fairly shaded from strong light on lower shore, in pools and in shallow water.

**Distribution** Atlantic, English Channel, North Sea and west Baltic. N.B. four varieties have been identified of which two are mentioned here.

### *Urticina eques* (Gosse)

Similar to *U. felina* (above) but rarely with debris adhering to column, larger with base up to 120mm or more in diameter, warts on column relatively inconspicuous; large tentacles reaching up to 60mm long; column usually higher than wide and not usually encrusted with gravel, etc.

**Habitat** on rocky substrates from 10–600m.

**Distribution** widespread and common; not recorded from the English Channel.

*Urticina eques*

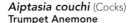

### Aiptasia couchi (Cocks)
**Trumpet Anemone**

Base lightly adhesive and sucker-like; orange-brown column with minute warts, broadening out towards the oral region and reaching 60mm high; sometimes column is patterned with cream-brown; 82 large tentacles, stout at the base and tapering to fine points, may be 30mm long and rarely retracted fully; could be confused with *Anemonia viridis* brown variety (see page 92) but characteristically has many pale lines running from the tentacle bases to the mouth, rather than the 2 white lines of *A. viridis*.

**Habitat** on rocks, extreme lower shore and shallow water; often local.

**Distribution** Mediterranean, Atlantic north to the English Channel approaches, Channel Islands, south Devon and Cornwall, South Wales.

*Aiptasia couchi*

### Haliplanella lineata (Verill)

Base adhesive, up to 20–50mm in diameter; column wrinkled when retracted, smooth and pillar-like when fully expanded; may reach up to 40mm tall, often less; up to 100 tentacles which may reach 100mm in length, they are retractile and delicate.

**Colour** dark green-brown with up to 20 vertical orange stripes; tentacles green-grey.

**Habitat** on rocks, stones, pebbles and shells, wooden piers etc, shallow water in sheltered bays and harbours.

**Distribution** Mediterranean, Atlantic, English Channel and North Sea.

*Haliplanella lineata*

### Diadumene cincta Stephenson

Base adhesive; smooth, orange-coloured column with characteristic collar at its upper limit (may not be seen when fully expanded), reaching to 30mm in height and 3.5mm in diameter; slender, light orange tentacles do not taper quickly, but are fully retractable; may be confused with the young of *Metridium senile*, orange variety (see page 96) but the tentacles of the latter species are less rapidly retracted.

*Diadumene cincta*

**Habitat** on rocks and shells on lower shore and in shallow water.

**Distribution** local and rare in English Channel and southern North Sea.

Check to see whether or not the specimen in question has blue spots on the periphery of the oral disc outside the tentacles, these are characteristic of *Actinia equina* (see page 92).

### Bunodactis verrucosa (Pennant)
**Gem Anemone**

Base adhesive, up to 30mm in diameter when the animal is closed; conical column reaches 30mm high and carries 6 rows of white warts interspersed with many more rows of blue-grey warts; about 48 transparent tentacles up to 15mm long.
**Colour** translucent and mottled green, grey or pink.
**Habitat** in rock pools on lower shore and in shallow water, frequently in small crevices well exposed to light and sometimes surrounded by sand.
**Distribution** Mediterranean, Atlantic, English Channel east to the Solent.

### Anthopleura balli (Cocks)

Base adhesive and sucker-like; warty column reaching to 50mm or more; warts largest near the top; column pink-orange or yellow-green which are the principal varieties of this variably coloured species (as illustrated); tentacles not readily retractile and translucent with grey, pink or brown mottlings and sometimes green tinges.
**Habitat** shuns strong light and occurs on rocks, particularly in holes or on the sides or undersides of crevices, on middle and lower shores and occasionally deeper.
**Distribution** Mediterranean, Atlantic north to the Irish Sea, English Channel east to Sussex and North Sea; often rare and local.

*Anthopleura thallia*

### Anthopleura thallia (Gosse)

Base lightly adhesive; column reaching up to 40mm with warts which are more pronounced near the top; column green-grey fading to the base; about 60 translucent; pale, tapering tentacles up to 25mm long.
**Habitat** middle and lower shore in pools on rocks, often in strong illumination (unlike *A. balli*);
sometimes gregarious
**Distribution** Atlantic north to south-west Scotland, English Channel east to Devon; rare.

*Bunodactis verrucosa*

*Anthopleura balli*
2 colour forms

95

### *Metridium senile* (Linnaeus)
**Plumose Anemone**

Base adhesive; column smooth, reaching up to 80mm or more; well-developed collar visible below the crown of fine, slender tentacles; tentacles are crowded and numerous, giving a 'feather duster' effect; tentacles generally more translucent than the column; form and shape of the individuals may be extremely variable, and the juveniles do not necessarily resemble the adults and can be difficult to identify.

**Colour** there are several colour varieties of which 2 are shown; others include brown, cream and orange.

**Habitat** on rocks, wrecks and piers from 0.5–3m.

**Distribution** Mediterranean, Atlantic, English Channel and North Sea.

### *Amphianthus dohrni* (von Koch)

Base elongated; about 20mm wide and modified to wrap around the stem of other organisms, e.g. the gorgonian *Eunicella verrucosa* (as shown here) or certain hydroids; individuals are sometimes closely grouped, thus giving a colonial effect, although they are really solitary; low, pillar-shaped column may be buff, pink or red and reaches up to 12mm in height; about 48 translucent, buff-white tentacles.

**Habitat** on rocky substrates from 20–600m wherever a suitable supporting organism can grow; this species cannot easily be confused with any other in the European area because of its habits.

**Distribution** Mediterranean, Atlantic north to the English Channel approaches and south-west Ireland.

*Metridium senile*

pink form

white form

*Amphianthus dohrni* on *Eunicella verrucosa*

### *Calliactis parasitica* (Couch)
#### 'Parasitic' Anemone

Base firmly adhesive; column stout and pillar-like, reaching upto 80mm; numerous tentacles in the adult, which may reach 30mm in length.
**Colour** column dark grey or brown with white or longitudinal stripes; tentacles translucent yellow-grey.
**Habitat** on muddy substrates from 3–100m.
**Distribution** Mediterranean, Atlantic north to the west of Ireland and the Irish Sea, and English Channel.
N.B. this is not a true parasite, but a commensal of the hermit crab. The anemone is frequently found attached to an empty shell of *Buccinum undatum* (see page 161). Several species of hermit crab may take up residence in the shell; the anemone protects the crab from the attacks of predators such as *Octopus,* and also obtains food from the crabs' feeding activities. Several anemones may be associated with 1 crab, but both crab and anemone can occur separately.

*Calliactis parasitica on shell occupied by Pagurus bernhardus*

### *Adamsia carcinopados* (Otto)

Commensal anemone almost always associated with the hermit crab *Pagurus prideauxi* (see page 231). Base and column highly modified to form an adhesive investment around the crab's body which itself is usually contained inside the shell of a small gastropod; as the crab grows the base of the anemone secretes a horny substance which effectively extends the shell; base may reach 70mm across if measured free from the crab; column is squat, giving way to tentacle disc almost immediately; about 500 tentacles.
**Colour** base brown to yellow, usually with red spots or blotches; tentacle disc and tentacles white and translucent, when disturbed the anemone may eject fine, lilac-purple threads (called acontia); may be immediately distinguished from *Calliactis parasitica* by the way the base is folded round the shell, as well as by the different colouring.
**Habitat** on sandy and muddy substrates from 4–100m.
**Distribution** Mediterranean, Atlantic, English Channel and North Sea.

*Adamsia carcinopados on shell occupied by Pagurus prideauxi*

*Sagartia elegans*
var. *venusta*

*Sagartia elegans*
var. *miniata*

*Sagartia ornata*

### Sagartia elegans (Dalyell)

Base strongly adhesive, up to 40mm in diameter but often less; red-brown column bearing wart-like suckers and reaching up to 60mm; up to 200 tentacles which may reach up to 15mm long.

**Habitat** on rocks and in crevices from the lower shore down to 50m; when disturbed the animal ejects white threads (acontia). N.B. there are several colour variants of which three are illustrated.

**1.** Variety *venusta* Tentacles and oral disc with no patterns; disc orange; tentacles white.

**Distribution** widely distributed around Ireland and the south-west of Britain.

**2.** Variety *rosea* (Not illustrated) As above but disc orange, white or drab; tentacles rose or magenta.

**Distribution** limited distribution in Ireland and south-west Britain.

**3.** Variety *nivea* (Not illustrated) As above, disc and tentacles white.

**Distribution** widely distributed around Ireland and the south and west of Britain.

**4.** Variety *miniata* Tentacles and mouth disc with well-marked pattern usually in shades of brown.

**Distribution** Atlantic, English Channel and North Sea. N.B. these varieties were formerly described as species in their own right by P. H. Gosse in his classical work of 1860, *Actinologia Britannica*.

### Sagartia troglodytes (Price)

Base firmly adhesive, diameter up to 35mm depending on the variety; column yellowish, becoming paler and greyer above with vertical lines; mouth disc and tentacles intricately patterned; with warty suckers reaching up to 40mm high; 100–200 tentacles, each up to 7.5mm long.

**Habitat** generally under stones or partly buried in mud or sandy mud and silt from lower shore down to 50m.

**Distribution** Atlantic, English Channel and North Sea north to Norway.

### Sagartia ornata (Holdsworth)

Base adhesive, column dark olive-green with slightly paler vertical lines and spots, wider towards base which may reach 15mm across, mouth disc and tentacles intricately patterned.

**Habitat** essentially a shore form inhabiting areas with cleaner sand than variety *S. troglodytes* (above).

**Distribution** the same, but more common on coasts on the south-west English Channel.

*Sagartia troglodytes*

### *Cereus pedunculatus* (Pennant)
**Daisy Anemone**
Base firmly adhesive, reaching 33mm in diameter; variably shaped, smooth column with grey suckers near the top, often trumpet-shaped when fully extended, and generally orange to buff-grey in colour; mouth disc often puckered and generally brown with darker markings; sometimes beautifully patterned; up to about 750 short tentacles of brown colour dotted and patterned with cream.
**Habitat** typically a shore species, in clean rock pools or in crevices as well as on stones and shells buried in mud (sometimes in great numbers); occasionally in estuaries.
**Distribution** Mediterranean, Atlantic north to Scottish borders and English Channel; not confirmed from the North Sea.

*Cereus pedunculatus*

### *Actinothoë sphyrodeta* (Gosse)
Base adhesive, reaching up to 10mm in diameter; grey-white, pillar-like column up to 30mm high; mouth disc sometimes coloured; may eject fine white threads when disturbed; may have over 100 grey-white tentacles.
**Habitat** on the undersides of pebbles or vertical and overhanging rockfaces and cave walls where the water is clean; on lower shore and in shallow water.
**Distribution** Atlantic to southwest Ireland and English Channel east to Dorset.

*Actinothoë sphyrodeta*

### *Sagartiogeton undata* (Müller)
Base firmly adhesive, reaching up to 40mm in diameter; column may contract to a flat mound with the upper part introverted, but in full expansion forms a tall pillar reaching up to 60mm; drab colour, yellower above with vertical, brown-grey stripes; may eject fine white threads when disturbed; oral disc light brown, bearing 100 slender, transparent tentacles.
**Habitat** normally attached to stones or shells buried in sandy mud on the extreme lower shore and down to 50m.
**Distribution** Mediterranean, Atlantic, English Channel and North Sea to Norway and west Baltic.

*Sagartiogeton undata*

## ORDER MADREPORARIA

### TRUE CORALS

Anthozoans with hard, calcareous skeletons into which the polyps can almost, if not completely, withdraw when disturbed. Often colonial. N.B. avoid confusion with calcified ectoprocts (see pages 244 to 249). Gorse, P. H. 1860 and Manuel, R. L. 1988 give more details.

*Caryophyllia smithi* with *Boscia anglicum*

### *Caryophyllia smithi* Stokes
#### Devonshire Cup Coral

Solitary, with stout, brown-white skeleton up to 15mm high and with conspicuous ridges (septa); polyps variable in colour – white, pink, brown or green, often with contrasting lips, e.g. red or green; tentacles variable in colour terminating in a small knob.

**Habitat** on rocks or stones from the extreme lower shore to 100m.

**Distribution** Atlantic, English Channel and northern North Sea (Scotland, Sweden, etc.). N.B. sometimes associated with the small barnacle *Megatrema anglicum* (see page 205) which may be found growing on the periphery of the coral cup, resembling a small wart 1–2mm high.

*Corynactis viridis*

skeleton of *Caryophyllia smithi* with *Boscia anglicum*

### *Balanophyllia regia* Gosse
#### Scarlet-and-gold Star Coral

Solitary, with cylindrical skeleton up to 10mm high; perforated septa not visible in life because polyps cannot withdraw entirely; brilliant orange or scarlet body; tentacles transparent with yellow flecks lacking a knob on the tip.

**Habitat** on extreme lower shore and in shallow water.

**Distribution** Atlantic, Bristol Channel and west English Channel; rare and local.

### *Lophelia pertusa* (Pallas)

Colonial, with irregularly branching, yellow-white skeletons reaching up to 500mm in height; pinkish polyps loosely scattered over the skeleton.

**Habitat** on rocky substrates in water from 60–600m.

*Balanophyllia regia*

Distribution Mediterranean (Adriatic), Atlantic (west coast of Scotland) and northern North Sea (west coasts of Sweden and Norway).

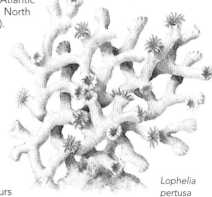

## ORDER CORALLIMORPHARIA

Anthozoans without a hard skeleton. The polyps have tentacles which terminate in a small knob.

### *Corynactis viridis* Allman
**Jewel 'Anemone'**
Solitary, small polyp with brilliant colours (which may fade in the aquarium); body 2.5–5mm in diameter; broad, adhesive base; tentacles arranged in 3 circlets; mouth borne on a minute cone.
Habitat on rocks on extreme lower shore and down to 100m.
Distribution Atlantic north to west English Channel and south-west Ireland.

*Lophelia pertusa*

## ORDER ALCYONACEA

### *SOFT CORALS*

Colonial anthozoans whose retractable polyps have 8 branching (pinnate) tentacles; polyps embedded in the body mass which has a skeleton made of a great number of free calcareous ossicles which makes these colonies soft and flexible. They are usually attached to rocks and stones.

*Alcyonium digitatum*

### *Alcyonium digitatum* (Linnaeus)
**Dead Man's Fingers**
Erect, stout, branching colonies up to 200mm high; white, yellow, orange or pink colour varieties; frequently branching in one plane and bearing a great number of white, retractable polyps which withdraw when disturbed; each polyp can extend up to 10mm: tentacles with about 16 branchlets on either side.
Habitat on rocks or stones on the lower shore and down to 100m.
Distribution Atlantic from Biscay northward, English Channel and North Sea.

### *Alcyonium palmatum* Pallas

Similar to *A. digitatum* (see page 101) but larger and more branched, reaching 500mm; colonies white, pink, brown or red with translucent white polyps; tentacles with 11–13 branchlets on either side.

**Habitat** attached to stones or shells on muddy substrates, or standing freely, in shallow water down to 20m.

**Distribution** Mediterranean, Atlantic to English Channel approaches, and south-west Britain.

*Alcyonium palmatum*

### *Parerythropodium coralloides* (Pallas)

Encrusting red colonies growing over branching stems of seaweeds or dead gorgonians (see below); polyps with white tentacles.

**Habitat** on rocky substrates where suitable supporting organisms grow, from 8–90m.

**Distribution** Mediterranean and Atlantic north to Biscay.

## ORDER GORGONACEA
## SEA-FANS AND THEIR ALLIES

Colonial anthozoans in which the polyps have 8 branching tentacles. The polyps are retractable, and embedded in tissue supported by a branching, central skeleton of calcium carbonate bound with a horn-like substance called *gorgonin*. They are anchored at the base by a holdfast.

*Parerythropodium coralloides* on dead *gorgonian*

### *Swiftia rosea* Madsen
**Small Sea Fan**

(Not illustrated) Colonial, reaching up to 150mm, smaller than *E. verrucosa* (below).

**Colour** white-grey.

**Habitat** below 10m deep, attached by holdfast to rock faces (usually sloping or vertical).

**Distribution** Local, Atlantic Scotland to Biscay.

### *Eunicella verrucosa* (Pallas)
**Sea-fan**

(Illustrated on page 96) Colonial, branching in one plane only, reaching 300mm in height; soft, pink outer tissue surrounds a

brown horny skeletal support; numerous pink polyps borne in slight swellings reaching up to 3mm when fully expanded.
**Habitat** attached by holdfast to rocks, usually from 15m downward.
**Distribution** Mediterranean, Atlantic north to west English Channel.

## ORDER PENNATULACEA
### SEA-PENS
Feather-shaped anthozoan colonies with a horny or chalky skeleton supporting the central column. The polyps are divided laterally, some-times on branches.

### *Pennatula phosphorea* Linnaeus
**Phosphorescent Sea-pen**
Feather-like colony with central column up to 400mm high; area bearing polyps roughly as long as that which does not.
**Colour** red; white polyps about 1mm long.
**Habitat** on sand and clay from 20m downward.
**Distribution** Mediterranean, Atlantic, North Sea and west Baltic.

*Pennatula phosphorea*

### *Virgularia mirabilis* (Muller)
Slender feather-like delicate erect colony reaching 200mm high; central column slender and when in situ, polyp bearing side branches grow right down to the surface of the substratum; side branches slender.
**Colour** cream-yellow.
**Habitat** on muddy bottoms from 30m.
**Distribution** local in harbours and sheltered places, Atlantic.

*Virgularia mirabilis*

### *Funiculina quadrangularis* (Pallas)
A large sea pen reaching up to 2m tall. The region bearing the polyps is flexible and a little more than half the total length; lower region, part of which is embedded in the substratum, relatively strong and stiff and slightly wide except for lance-like tapering tip; two forms of polyps branch from upper region.
**Colour** pink.
**Habitat** on muddy bottoms below 20m.
**Distribution** Atlantic.

*Funiculina quadrangularis*

103

These animals differ in several respects from the cnidarians and thus are often separately classified. The body may be variously shaped and composed of two thin layers of cells separated by a volume of transparent, iridescent or luminous jelly which constitutes the animal's main bulk. The mouth is situated at the bottom of the body and leads into a series of digestive canals which open by one or two minute pores at the top. Radiating from the body are a series of up to eight swimming structures called *comb-rows*. Each comb-row is made up of a number of plates consisting of fused cilia. These beat up and down rhythmically, so driving the animal through the water.

*Pleurobrachia pileus*

Ctenophores are highly predatory and feed on other floating animals. Many have tentacles which can be protruded from pits on either side of the body and trailed along like fishing lines. These tentacles carry cells which cannot sting the prey, but lasso it, thus securing it until it is passed to the mouth. More information is given by Newell, G. E. and Newell, R. C. 1973.

## CLASS TENTACULATA
### Ctenophores with retractile tentacles

**Pleurobrachia pileus** (O. F. Müller)
**Sea-gooseberry**
Rounded, oval body up to 30mm in length with conspicuous comb-rows running from the apex, but terminating short of the bottom of the animal; relatively long, branching tentacles.
**Colour** white-orange gut.
**Habitat** common in open water and occasionally found in rock pools; may be in shoals.
**Distribution** Mediterranean, Atlantic, English Channel, North Sea and Baltic.

**Bolinopsis infundibulum** (O. F. Müller)
Oval body reaching up to 150mm in length; 2 conspicuous comb-rows at either side of the mouth which may be half the length of the rest of the body; branched tentacles.
**Habitat** open water; sometimes in shoals.
**Distribution** Mediterranean, Atlantic, English Channel, North Sea and west Baltic.

**Cestus veneris** Lesueur
Ribbon-like body, 80mm long and 150mm high; 4 transparent comb-rows; main tentacles reduced, but secondary tentacles lie in 2 grooves near the mouth.
**Colour** transparent, sometimes green-violet.
**Habitat** open water.
**Distribution** Mediterranean and Atlantic usually south of Britain; extremely rare in North Sea.

## CLASS NUDA
Ctenophores lacking tentacles

**Beroë cucumis** Fabricius
Mitre-shaped body up to 100mm or more long; comb-rows run from the apex to the base; branched inner canals visible.
**Colour** transparent; occasionally pinkish.
**Habitat** open water.
**Distribution** Mediterranean, Atlantic, North Sea and Baltic.

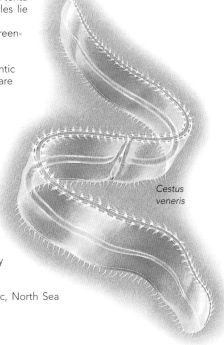

Cestus
veneris

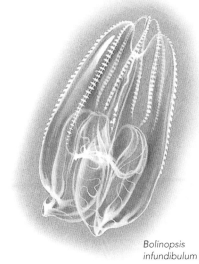

Bolinopsis
infundibulum

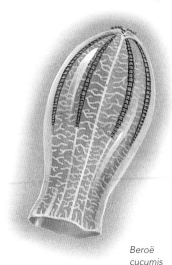

Beroë
cucumis

105

## Marine worms

The animal kingdom includes many different types of worms. The different types cover a range of body plans and life styles and are grouped as follows; phylum Platyhelminthes (the flatworms); phylum Nemertina (the ribbon worms); phylum Nematoda (the round worms); phylum Annelida (the segmented worms); and phylum Priapuloidea, phylum Echiuroidea, phylum Sipunculoidea (these last three phyla are generally regarded as minor groups).

Apart from their bilateral symmetry, bodies composed of three cell layers, and the fact that they all require a moist if not aquatic environment, these groups have relatively little in common. Disposition of appendages, presence or absence of segmentation, number of body openings, habit and the pattern of locomotion should all be of assistance when seeking to distinguish between these various phyla.

## Class Turbellaria

Generally free-living, leaf-shaped worms. They lack a body cavity separating the gut from the remaining tissues. The mouth opens on the underside, and the pharynx is often everted for feeding. The gut is simple or branched, and sometimes visible through the skin; there is no anus. Rudimentary sense organs at the anterior end include eye spots and tentacles. Locomotion is with a characteristic, gliding movement effected by the combined action of thousands of cilia on the underside, although the animals can change their shape by muscle contraction. A further account of Baltic turbellarians may be obtained from Forsman, B. 1972 and Ball, I. R. and Reynoldson, T. B. 1981 and Prudhoe, S. 1982 give excellent accounts of many members of this group.

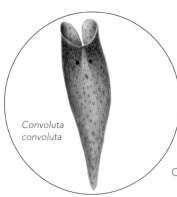

*Convoluta convoluta* (Abildgaard)
  **Length** up to 3mm.
    **Head** broader than tail and lacking distinct tentacles.
    **Body** flattened and leaf-like or pear-shaped, often turned under at the edges; gut lacking.
    **Colour** brown to green, due to symbiotic algae.
    **Habitat** among seaweeds from lower shore down to 15m.
  **Distribution** Mediterranean, Atlantic, English Channel, North Sea and Baltic.

### *Monocelis lineata* (O. F. Müller)
**Length** up to 2mm.
**Head** not readily discernible from body and lacking distinct tentacles.
**Body** slightly pointed at the head end; gut is simple.
**Colour** opaque white gut visible through skin.
**Habitat** among seaweeds (e.g. *Ulva* see page 23) from middle shore downward.
**Distribution** Mediterranean, Atlantic, English Channel and North Sea.

### *Procerodes littoralis* (Strom)
**Length** up to 5mm.
**Head** bears triangular tentacles and 2 eyes.
**Body** rounded at the rear and tapering towards the head; gut with 3 branches.
**Habitat** under stones, often near freshwater outlets, from upper shore downward.
**Distribution** Atlantic, English Channel, North Sea and Baltic.

*Monocelis
lineata*

*Procerodes
littoralis*

### *Oligocladus sanguinolentus* (Quatrefages)
**Length** up to 13mm.
**Head** rounded, bears 2 tentacles and many eyes.
**Body** flattened and leaf-shaped with rounded tail; gut visible and with a number of branches.
**Colour** transparent body, white above with brown-red spots.
**Habitat** under stones and seaweeds from middle shore and downward.
**Distribution** Atlantic, English Channel and North Sea.

*Oligocladus
sanguinolentus*

### *Prostheceraeus vittatus*
(Montagu)
**Length** up to 30mm or more.
**Head** bluntish, bears 2 conspicuous tentacles.
**Body** flat, leaf-shaped, with edges thrown into folds and tapering to a point at the tail.
**Habitat** under stones in mud.
**Distribution** Atlantic, western English Channel and North Sea.

*Prostheceraeus
vittatus*

# PHYLUM NEMERTINA RIBBON WORMS

These ribbon-shaped worms are often extremely long. The unsegmented body is composed of three cell layers and there is no body cavity separating the gut from the other tissues. The mouth is anterior and the anus posterior, with a characteristic proboscis opening via the mouth to capture and handle prey. Rudimentary sense organs including eyespots are present.

Nemertines are frequently abundant animals but because they are fragile and often burrow in sand and mud, they may be overlooked. Many aspects of the general biology are well described by Gibson, R. 1972. The extension of the proboscis may greatly increase the apparent length of the animal. The form and disposition of the eyes, together with the shape of the head, are most helpful in assisting identification, and there is generally a slit along each side of the head which should also be looked for with the assistance of a hand lens. For this reason the heads of the animals have been illustrated approximately one and a half times larger than their bodies. For identification of other nemertines see McIntosh, W. C. 1873–1923, and Brunberg, L. 1964. Gibson, R. 1982 gives an up to date account for the identification of these worms.

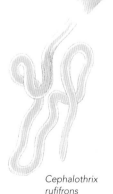

### Tubulanus annulatus (Montagu)
**Length** up to 120mm, occasionally 700mm.
**Body** flat below, rounded above, narrowing behind the head and tapering towards the tail; no eyes visible, but head slits open just behind the snout.

*Tubulanus annulatus*

**Colour** head usually paler than patterned body.
**Habitat** in sand under stones, in rocky clefts or in annelid worm tubes from lower shore down to 10m and beyond.
**Distribution** Mediterranean, Atlantic, English Channel and North Sea.

### Cephalothrix rufifrons (Johnston)
**Length** up to 80mm, occasionally more.
**Body** shape variable, according to state of extension, generally tapering towards head and tail; adult bears no eyes or other marks on head.
**Colour** generally pale yellow with red-orange head; often a distinct line runs part way down the centre of the body.
**Habitat** in coarse, clean sand, in pools amongst coralline algae, under stones and among shells from the lower shore downward.
**Distribution** Atlantic, English Channel and North Sea.

### Cerebratulus fuscus (McIntosh)
**Length** up to 150mm.
**Head** bears 4–8 eyes with deep slits.
**Body** flattened and tapering towards head and tail, tail bearing a thin, terminal filament.
**Colour** varies from skin colour to grey-brown.
**Habitat** among coralline seaweeds, shells or pebbles associated with Laminaria holdfasts

*Cephalothrix rufifrons*

*Cerebratulus fuscus*

(see page 32–33), or in mud, from lower shore down to 100m or more.
**Distribution** Mediterranean, Atlantic and English Channel.

### Lineus longissimus (Gunnerus)
**Length** up to 5m, but much larger specimens have been recorded.
**Head** bears a dense group of eyes on each side, and deep slits behind.
**Body** rounded, narrowed behind the head and tapering for the last part only.
**Colour** brown-ivory.
**Habitat** under stones and among rocks from lower shore and in shallow water.
**Distribution** Atlantic, English Channel, North Sea and the west Baltic.

*Lineus longissimus*

### Lineus bilineatus (Renier)
**Length** up to about 700mm.
**Head** broad, lacks eyes but bears deep slits.
**Body** tapering towards tail.
**Colour** shades of brown, with white line down the middle of the back.
**Habitat** in deeper water among coralline seaweeds and shells, occasionally under shells on lower shore.
**Distribution** Mediterranean, Atlantic, English Channel and North Sea.

*Lineus bilineatus*

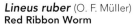

### Lineus ruber (O. F. Müller)
**Red Ribbon Worm**
**Length** up to 80mm, seldom longer.
**Head** spatulate, fractionally wider than the adjoining part of the body with shallow slits; 3 or 4 eyes in a row on each side.
**Body** flattened, the latter part tapering towards the tail.
**Colour** red-brown; ventral surface paler than dorsal surface.
**Habitat** under stones among muddy gravel from middle to lower shore down to deep water.
**Distribution** Mediterranean, Atlantic. English Channel, North Sea and Baltic.

*Lineus ruber*

### Micrura aurantiaca (Grube)
**Length** up to 100mm.
**Head** bears short, white snout which lacks eyes; mouth has shallow slits.
**Body** flat below and rounded above.
**Colour** brick-red with white proboscis.
**Habitat** under stones in rock pools and in deeper water.
**Distribution** Mediterranean, Atlantic and English Channel.

*Micrura aurantiaca*

### *Prosorhochmus claparèdi* (Keferstein)

*Prosorhochmus claparèdi*

**Length** up to 35mm.

**Head** broad and spatulate, often slightly wider than the body and with a conspicuous central notch from which a pale streak leads beyond the grey ganglia; 4 eyes situated well back, the anterior pair larger than the posterior pair and the left eyes widely separated from the right; 1 pair of slits.

**Body** flattened, with a slight constriction behind the head, tapering in the region of the tail.

**Colour** pale yellow but occasionally orange.

**Habitat** in crevices and among fissures in rocks on upper and middle shore.

**Distribution** Atlantic and English Channel.

### *Oerstedia dorsalis* (Abildgaard)

**Length** up to 25mm but often less.

**Head** slightly notched in front; 4 eyes arranged in a square; 1 pair of slits.

**Body** almost circular in cross section and slightly tapering at each end.

**Colour** dorsal surface brown-red with either yellow granules or a yellow dorsal stripe, and ventral surface paler; or green-brown with brown annular markings and a white stripe.

**Habitat** among laminarians and other growths down to 20m or more, seldom on the shore.

**Distribution** Mediterranean, Atlantic, English Channel, North Sea and west Baltic.

*Oerstedia dorsalis*

2 colour forms

### *Amphiporus lactifloreus* (Johnston)

**Length** up to 100mm but often less.

**Head** flat and spatulate; several groups of eyes arranged in a marginal row on each side, and in groups over or close to the conspicuous pink ganglia, giving an overall near-triangle of eyes; 2 sets of slits run obliquely.

*Amphiporus lactifloreus*

**Body** flattened ventrally and rounded dorsally, and not tapering towards the blunt tail.

**Colour** various shades from pink to white, but pink predominates; translucent line along dorsal surface indicates the position of the retracted proboscis.

**Habitat** under stones on the lower shore and associated with laminarians, and in deeper water.

**Distribution** Mediterranean, Atlantic, English Channel, North Sea and west Baltic.

### *Prostomatella obscurum* (Schultze)
### *( =Prostoma obscurum)*

**Length** up to 30mm.
**Head** bears 4 eyes.
**Body** relatively short.
**Colour** grey-green with a longitudinal green stripe down the middle of the back.
**Habitat** among seaweeds and mud in areas of reduced salinity, generally in shallow water; this is a brackish water species.
**Distribution** Baltic.

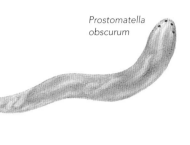

*Prostomatella obscurum*

### *Tetrastemma melanocephalum*
(Johnston)

**Length** up to 60mm.
**Head** flattened and generally slightly wider than the body, with a conspicuous frontal notch; 4 eyes, the first pair of which lie within the conspicuous patch of black skin so that they are not readily distinguished, while the second pair lie behind the pigment patch and are conspicuous; 1 pair of obliquely set slits.
**Body** flattened when extended but more rounded when contracted, slightly constricted behind the head and not tapering until near the tail.
**Colour** dull yellow-green, with a squarish black patch on the head.
**Habitat** on seaweeds and under stones on the lower shore and down to 60m.
**Distribution** Mediterranean, Atlantic, English Channel, North Sea and west Baltic.

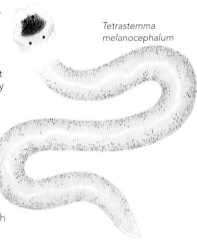

*Tetrastemma melanocephalum*

### *Punnettia splendida* (Keferstein)

**Length** up to 80mm.
**Head** conical or diamond-shaped; many eyes situated dorsally and laterally; oblique slits with accessory furrows.
**Body** flattened, thickening suddenly behind the head.
**Colour** red-yellow with 5 conspicuous, ivory, longitudinal stripes.
**Habitat** under stones, in rock crevices and among shells down to 40m.
**Distribution** Mediterranean, Atlantic and English Channel.

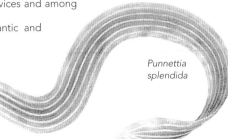

*Punnettia splendida*

111

# PHYLUM ANNELIDA SEGMENTED WORMS

A very important group of worms with an estimated 14,000 species. The body is developed from three cell layers and encloses a fluid-filled body cavity (the coelom). The body is divided lengthwise into a number of recognizable segments, each one of which usually carries bristles known as *chaetae* as well as other structures. The head is often well developed, bearing sense organs and a simple brain, and the mouth opens by the second segment (the first segment is the prostomium or preoral segment). The anus opens on the terminal segment. Well-developed longitudinal and circular muscles of the body wall allow extension and contraction of the body, and are often associated with locomotory segmental appendages.

This phylum is divided into three large classes; the bristle worms (Polychaeta), the earthworms and their allies (Oligochaeta) and the leeches (Hirudinea). Of these, the polychaetes have representatives in almost all marine environments, but they rarely occur in fresh water and only very rarely in damp terrestrial habitats. The other classes have only a few marine representatives. Something of the importance of the polychaetes will be appreciated by readers exploring the sea and the shore because of the abundance, diversity and the exceptionally wide geographical distribution of these animals.

Throughout the accompanying diagrams, the following key letters have been applied: ac=acicula (rod which supports the parapodium); an=antenna; c=cirrus (small outgrowth of head or parapodium); ch=chaeta, bristle or seta; d=dorsal surface; e=eye; ft=free tooth or paragnath; g=gill; lo=lateral organ (small structure on parapodium of some species); j=jaw; mi=marginal membrane of head of some species, membrane is incised at the edge; ms=marginal membrane of head of some species, membrane has a scalloped edge; p=palp; pr=prostomial process; pro=proboscis; s=scale; t=tooth; tc=tentacular cirrus of head; v=ventral surface.

## CLASS POLYCHAETA Bristle worms

Annelids with a preoral segment or prostomium. The head is formed from several highly modified segments fused together and carries a variety of specialized structures such as antennae, eyes, palps, jaws and tentacular cirri (see fig. 1). The rest of the body is composed of a relatively large number of similar segments, most of which bear a pair of locomotory appendages called parapodia; these sometimes have a respiratory function also. The parapodia are composed of several structures which are important in the identification of the various species (see fig. 2). The sexes are usually separate, and fertilization takes place in the sea; a pelagic larva is often formed. Habits and habitats are very variable. These annelids bear many chaetae (bristles) on each segment, as opposed to the oligochaetes (earthworms and their allies) which bear a few chaetae on each segment.

This diverse group of worms cannot be divided satisfactorily into orders, but falls into approximately 80 families of which twenty-six are treated in this book. The basic body plan is modified in the various groups The poly-

▼ Fig. 1 Head of *Lepidonotus clava*

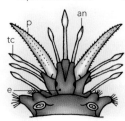

chaetes may be described loosely as errant (free living and predacious) or sedentary (burrowing or tube-dwelling). Three characteristics in particular help to identify the worms correctly. The presence or absence of a tube and the form of the tube itself is a clue to a number of famlies. A hard, calcareous tube occurs in the Serpulidae and Spirorbidae, whereas in the Pectinariidae ( =Amphictenidae) and many Terebellidae the tube is of sand cemented into an organic matrix; and in the Sabellidae the tube is built of mud. If the worm is a burrowing variety then the shape and form of the burrow should be considered. Next, the form of the head and the disposition of the appendages should be examined. These are shown in fig. 1 and in a number of other cases on the following pages. Finally, the form of the parapodia should be investigated. A hand lens may be essential, and mounting parapodia on a slide and examining them with a microscope will often assist further. Fig. 3 shows the arrangement of bristles or chaetae as well as the dorsal and ventral parapodial cirri and other structures in a typical polychaete.

When examining an unidentified worm it should be remembered that errant polychaetes generally have well-developed eyes and tentacles for receiving information from the environment and for seeking prey. They often have a protrusible proboscis which may be armed with powerful jaws, and their parapodia are powerfully built for crawling or swimming. Generally the form of the proboscis may be discerned only when it has been everted. Pressing the pharyngeal region of such worms either when they are alive or narcotized may cause the proboscis to be everted. In other proboscis-bearing species the shape of the organ can sometimes be made out through the relatively transparent body wall of the anterior region. Fig. 4 shows the general characters of an errant polychaete's anterior end. Sedentary polychaetes usually have reduced sensory structures and reduced parapodia but their gills may be large and conspicuous, and are often also used to filter food from the water as well as to extract oxygen.

Fauvel, P. 1923 and 1927 provides a detailed identification for most of the European polychaetes, while Fauchald, K. 1977 gives keys to the orders, families and genera. George, J. G. and Hartman-Schröder, G. 1985 and Chambers, S. J. and Muir, R. I. 1997 deal in detail with some of the families covered here. McIntosh, W. C. 1873–1923 is very out of date taxonomically but has some wonderful illustrations.

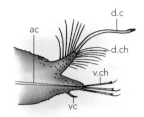

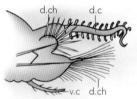

▲ **Fig. 2** Cirrus-bearing (top), and scale-bearing parapodia of *Hermione hystrix*

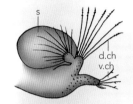

▲ **Fig. 3** Parapodium of *Harmothoë impar*

▼ **Fig. 4** Anterior end of male *Exogone naidina*

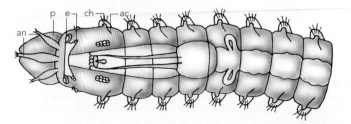

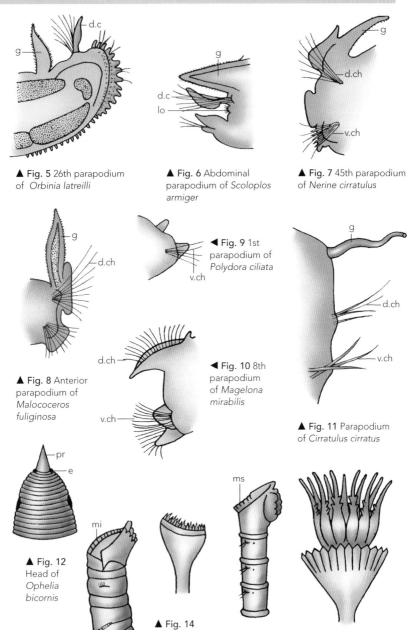

▲ Fig. 5 26th parapodium of *Orbinia latreilli*

▲ Fig. 6 Abdominal parapodium of *Scoloplos armiger*

▲ Fig. 7 45th parapodium of *Nerine cirratulus*

◀ Fig. 9 1st parapodium of *Polydora ciliata*

▲ Fig. 8 Anterior parapodium of *Malococeros fuliginosa*

◀ Fig. 10 8th parapodium of *Magelona mirabilis*

▲ Fig. 11 Parapodium of *Cirratulus cirratus*

▲ Fig. 12 Head of *Ophelia bicornis*

▶ Fig. 13 Head of *Maldane sarsi*

▲ Fig. 14 Operculum of *Mercierella enigmatica*

▲ Fig. 15 Head of *Euclymene lumbricoides*

▲ Fig. 16 Operculum of *Hydroides norvegica*

114

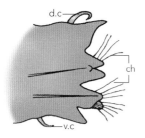

▲ Fig. 17 16th parapodium of *Nereis diversicolor*

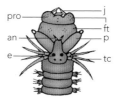

▲ Fig. 18 Head of *Nereis pelagicar* (proboscis everted)

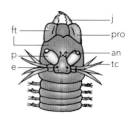

▲ Fig. 19 Head of *Nereis diversicolor* (proboscivs everted)

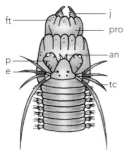

▲ Fig. 20 Head of *Nereis virens* (proboscis everted)

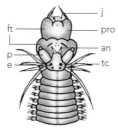

▲ Fig. 21 Head of *Perinereis cultrifera* (proboscis everted)

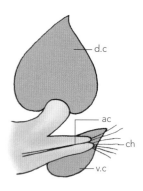

▲ Fig. 22 Parapodium of *Phyllodoce paretti*

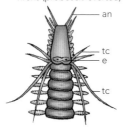

▲ Fig. 23 Head of *Phyllodoce lamelligera*

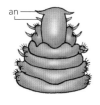

▲ Fig. 24 Head of *Nephtys caeca*

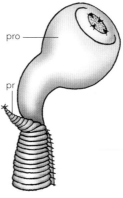

▲ Fig. 25 Head of *Glycera convoluta* (proboscis everted)

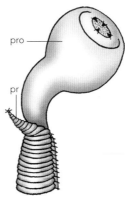

▲ Fig. 26 Head of *Eunice harassii*

## Family Aphroditidae Scale worms

Dorso-ventrally flattened, free-living polychaetes, dorsal surface is partly or entirely covered by interfolded scales. Head bears one pair of antennae.

### *Aphrodite aculeata* Linnaeus
#### Sea-mouse

**Length** 100–200mm. Readily distinguished from other members of the family by its large bulk and mat of grey-brown hairs masking the scales on the dorsal surface.

**Body** oval shaped; convex above, flat and sole-like below; about 40 segments bear conspicuous, gold-brown or iridescent chaetae on the ventral surface.

**Habitat** on soft substrates in shallow and deeper water.

**Distribution** Mediterranean, Atlantic, English Channel, North Sea and west Baltic.

## Family Polynoidae

Dorso-ventrally flattened polychaetes, dorsal surface partly or entirely covered with scales. Head bears up to 3 antennae, see fig. 1, page 112. N.B. handling may dislodge scales so giving a false impression of their numbers.

*Aphrodite
aculeata*

*Hermione
hystrix*

### *Hermione hystrix* (Savigny)

**Length** up to 60mm.

**Head** hidden from above, bears 2 eyes, 1 long antenna, 2 long palps and 2 pairs of tentacular cirri.

**Body** flat and oval shaped; dorsal surface bears scales which overlap left and right; 32–34 segments bear chaetae; parapodia of two types, bearing cirri or scales (see fig. 2, page 113) alternating with each other.

**Colour** red-brown above.

**Habitat** in gravel, sand and mud, and among bivalves such as scallops and oysters down to about 100m.

**Distribution** Mediterranean, Atlantic, English Channel and North Sea.

### *Lepidonotus clava* (Montagu)

**Length** up to 30mm.

**Head** partly hidden from above, bears appendages as shown in fig. 1, page 112.

**Body** flat, covered for most of its length by rounded non-overlapping scales; 26 segments bear chaetae arranged in two groups per parapodium; alternate parapodia bear scales; dorsal cirrus recurved.

*Lepidonotus
clava*

Colour brown.
Habitat under rocks and stones on the lower shore.
Distribution Mediterranean, Atlantic and English Channel.

### Harmothoë impar (Johnston)
Length up to 25mm.
Head hidden from above, bears 4 eyes, 1 median and 2 lateral antennae, 2 stout palps and 2 pairs of tentacular cirri.
Body flat, completely covered by 15 pairs of overlapping scales which carry papillae on their lateral edges; 35–40 segments bear chaetae; form of parapodia shown in fig. 3, page 113; alternate parapodia bear scales.
Colour brown-green scales with a yellow-grey central spot.
Habitat under stones, rocks and seaweeds on the lower shore and in shallow water.
Distribution Mediterranean, Atlantic, English Channel, North Sea and west Baltic. N.B. Harmothoë is a large genus so there are 20 similar species.

### Scalisetosus assimilis (McIntosh)
(Not illustrated)
Length up to 20mm.
Head not entirely covered by 15 pairs of transparent scales.
Habitat among the spines of Echinus esculentus (see page 263).
Distribution Atlantic, English Channel and North Sea.

Harmothoë
impar

### Polynoë scolopendrina (Savigny)
Length up to 120mm.
Body has 15 pairs of overlapping scales which cover about half the length of the body; 80–100 segments bear chaetae; alternate parapodia from segment 2 to 32 carry scales.
Colour variable; head is often red, as may be the body below the metallic-sheened scales.
Habitat in other worm tubes, cracks in rock and in sand.
Distribution Mediterranean, Atlantic, English Channel and North Sea.

### Acholoë astericola (Delle Chiaje)
(Not illustrated)
Length up to 50mm.
Colour male opaque white and female orange-red.
Habitat always in the ambulacral groove of the starfishes Astropecten irregularis and Luidia ciliaris (see pages 252–253).
Distribution Mediterranean, Atlantic and English Channel.

Polynoë
scolopendrina

117

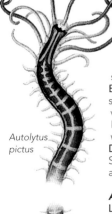

detail of
dorsal cirrus

*Syllis
prolifera*

*Autolytus
pictus*

## Family Syllidae

Small, delicate, free-living polychaetes with thread-like bodies, often beautifully coloured. The head generally possesses 4 eyes, 3 antennae, 2 palps, and 2 pairs of tentacular cirri. The eversible proboscis is divided into two parts, of which the anterior appears cylindrical, chitinous and armed with 1 or more teeth. Parapodia are not bilobed, and dorsal and ventral cirri are usually present (the dorsal cirri are often long, 'jointed' and conspicuous); 2 anal cirri are present. Reproduction is sometimes by budding, so that stolons of several individuals arranged in a chain may be encountered. This is a large family which includes many genera and species, of which only a very few can be included here.

### *Syllis prolifera* Krohn

**Length** 10–25mm.

**Head** bears proboscis with 1 tooth; 4 eyes; 2 minute ocelli; middle antennae longer than lateral antennae; palps slightly triangular.

**Body** dorsal cirri of parapodia long, and composed of 20–40 'joints'.

**Colour** body very variable, greyish to reddish, anterior region sometimes with brown, pink or orange markings.

**Habitat** on the lower shore and in shallow water among sea-weeds and stones.

**Distribution** Mediterranean, Atlantic, English Channel and North Sea.

### *Exogone naidina* (Oersted)

(See fig. 4, page 113)

**Length** up to 4mm.

**Head** bears 4 large eyes; 3 antennae arranged in a line in front of the eyes, with the middle antenna longer than the lateral antennae; 1 pair of tentacular cirri reduced to small, button-like processes; form of head shown in fig. 4, page 113.

**Body** 24–33 segments bear chaetae; dorsal cirri ovoid and shorter than the parapodium which bears them, ventral cirri very small, 2 long anal cirri.

**Habitat** on the lower shore and in shallow water among sea-weeds, bryozoans and ascidians.

**Distribution** Mediterranean, Atlantic, English Channel, North Sea and west Baltic. N.B. eggs and embryos may be found attached to the underside of the female; eggs reddish in colour.

### *Autolytus pictus* (Ehlers)

**Length** up to 25mm.

**Head** bears proboscis with 10 large teeth, alternating with 10 smaller teeth; if withdrawn, proboscis may be seen as an s-shaped organ; longest tentacular cirri almost as long as lateral antennae.

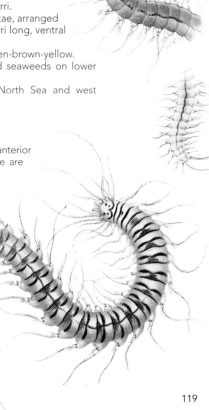

*Kefersteinia cirrata*

Body 60–100 segments bear chaetae; form of parapodia varies along the body according to function, dorsal cirri of 1st chaeta-bearing segment are very long and conspicuous while those on the 2nd are much shorter; 2 thick anal cirri.

Colour variable; pale-pink below, violet markings above.
Habitat among laminarian holdfasts (see page 32–33), sponges and stones.
Distribution Mediterranean, Atlantic, English Channel and North Sea.

## Family Hesionidae
Free-living polychaetes with cylindrical bodies, and segmentation weakly marked. The eversible proboscis may bear jaws. Parapodia bear long dorsal cirri.

### *Kefersteinia cirrata* (Keferstein)
Length up to 75mm.
Head bears a short, thick, eversible proboscis without jaws, which has a large opening surrounded by many papillae; 4 eyes; 2 palps thicker than the 2 smaller antennae; 8 pairs of tentacular cirri.
Body fragile; 36–65 segments bear chaetae, arranged in one group per parapodium; dorsal cirri long, ventral cirri short.
Colour varies with sex and maturity; green-brown-yellow.
Habitat among worm tubes, shells and seaweeds on lower shore and in shallow water.
Distribution Mediterranean, Atlantic, North Sea and west Baltic.

### *Castalia punctata* (O. F. Müller)
Length up to 25mm.
Head bears eversible proboscis; 4 eyes, anterior larger than posterior; 2 lateral antennae are thinner than the 2 palps; 6 pairs of long tentacular cirri.
Body 40–50 segments bear chaetae; dorsal cirri much longer than ventral cirri.
Habitat among stones, shells and seaweeds from lower shore downward.
Distribution Atlantic and North Sea.

*Castalia punctata*

### Family Nereididae

*Nereis pelagica*

Often large, active, free-living polychaetes. The head bears an eversible proboscis armed with large, black jaws and accessory "free" teeth sometimes called paragnaths (see fig. 18, 19 and 20, page 115), 4 eyes, 2 antennae, 2 ovoid palps terminating with small button-like structures and 4 pairs of tentacular cirri. The bodies possess many segments bearing chaetae on well-developed parapodia (see fig. 17, page 115).

### *Nereis pelagica* Linnaeus

**Length** up to 120mm.

**Head** proboscis carries free teeth as well as the 5–7 teeth on each of the jaws; form of head shown in fig. 18, page 115.

**Body** cylindrical, tapering posteriorly; 80–100 segments bear chaetae; dorsal cirrus of parapodia long and fairly conspicuous.

**Colour** adult normally red-brown-yellow with a distinct dorsal blood vessel.

**Habitat** among rocks, shells and seaweeds on lower shore, and in shallow water.

**Distribution** Mediterranean, Atlantic, English Channel, North Sea and west Baltic.
N.B. a pelagic, reproductive variety (known as the epitoke) may be encountered, whose posterior segments carry parapodia modified for swimming.

*N. pelagica* epitoke

*Nereis diversicolor*

### *Nereis diversicolor* (O. F. Müller)
**Rag Worm**

**Length** up to 120mm.

**Head** bears a proboscis carrying free teeth as well as the 5–8 teeth on each of the jaws; posterior tentacular cirri extend back as far as the 5th or 7th chaeta-bearing segments; form of head shown in fig. 19, page 115.

**Body** 90–120 segments bear chaetae.

**Colour** very variable, but often green-yellow with tints of orange and red; distinct dorsal blood vessel makes a red line all the way down the back.

**Habitat** from the middle shore down to shallow water, burrowing in sand or mud, often in brackish conditions.

**Distribution** Mediterranean, Atlantic, English Channel, North Sea and west Baltic.

### *Nereis virens* M. Sars
**King Rag Worm**
**Length** usually up to 200mm, may occasionally reach 400mm.
**Head** bears proboscis carrying free teeth as well as the 6–10 teeth on each of the jaws; posterior tentacular cirri reach back as far as the 5th or 8th chaeta-bearing segment; form of head shown in fig. 20, page 115.
**Body** large, may be as thick as the average finger; 100–175 segments bear chaetae; parapodia of complex form with many appendages varying in proportion along the body.
**Colour** adult green with iridescent purple, edges of parapodia bordered with yellow.
**Habitat** on lower shore and in shallow water, often burrowing in sand.
**Distribution** Atlantic around the north and west coasts of Britain and Ireland, North Sea and west Baltic. N.B. a pelagic epitoke with a slightly modified body may be encountered.

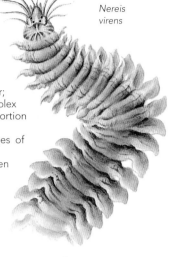

*Nereis virens*

### *Nereis fucata* (Savigny)
**Length** up to 200mm.
**Head** posterior tentacular cirri extend back as far as the 3rd or 5th chaeta-bearing segment.
**Body** 90–120 segments bear chaetae.
**Colour** adult brown-yellow with white markings in the middle of each segment.
**Habitat** the adults are found inside whelk shells (often of *Buccinum undatum*, see page 161) which are occupied by hermit crabs (e.g. *Pagurus bernhardus* and *P. prideauxi*, see page 231).
**Distribution** Mediterranean, Atlantic, English Channel and North Sea. N.B. a pelagic epitoke may be encountered.

### *Perinereis cultrifera* (Grube)
**Length** up to 250mm.
**Head** posterior tentacular cirri extend back as far as the 5th or 6th chaeta-bearing segment; form of head shown in fig. 21, page 115.
**Body** somewhat flattened and tapering towards the tail; 100–125 segments bear chaetae.
**Colour** adult brown-green with parapodia reddish above.
**Habitat** on gravel, sand or mud on lower shore and in pools and shallow water.
**Distribution** Atlantic, English Channel and North Sea. N.B. a pelagic epitoke may be encountered.

*Nereis fucata*

*Perinereis cultrifera*

*Phyllodoce
lamelligera*

## Family **Phyllodocidae** Paddle worms

Free-living polychaetes with parapodia whose dorsal cirri are typically large, conspicuous and leaf-like, so giving the appearance of paddles. The ventral cirri are similar but smaller, and the chaetae are generally arranged in one group per parapodium (see fig. 22, page 115); 2 anal cirri.

### *Phyllodoce lamelligera* (Gmelin)

**Length** from 60–600mm.
**Head** bears an eversible proboscis which lacks jaws; 2 black eyes; 4 antennae; 4 pairs of tentacular cirri; form of head shown in fig. 23, page 115.
**Body** 300–400 segments; parapodia bear conspicuous, lance-shaped olive or green dorsal cirri.
**Colour** blue-brown.
**Habitat** under rocks and stones from the lower shore downward, often among laminarians.
**Distribution** Mediterranean, Atlantic and English Channel.

### *Phyllodoce paretti* (Blainville)
**Paddle Worm**

**Length** 150–300mm.
**Head** bears an eversible proboscis which lacks jaws; 2 large eyes; 4 short antennae; 4 tentacular cirri which may not all be readily visible from above.
**Body** long and tapering at either end; about 200 segments, form of parapodia shown in fig. 22, page 115.
**Colour** variable, often dark blue above, with black, green or yellow markings on parapodia.
**Habitat** under stones and rocks by day, often in sandy and muddy places, from the lower shore downward.
**Distribution** Mediterranean, Atlantic, English Channel and North Sea.

### *Phyllodoce maculata* (Linnaeus)

Generally similar to *P. paretti* (above).
**Length** up to 100mm.
**Head** bears eversible proboscis which lacks jaws.
**Body** about 250 segments with chaetae; conspicuous, brown-grey, leaf-like dorsal cirrus on parapodia.
**Colour** green-yellow with three or four brown patches on the back of each segment.
**Habitat** in sand and mud under pebbles, lower shore downward.
**Distribution** Atlantic, English Channel, North Sea and west Baltic.

*Phyllodoce
paretti*

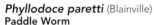

*Phyllodoce
maculata*

### *Eulalia viridis* (O. F. Müller)
**Green Leaf Worm**
**Length** 50–150mm.
**Head** small, rounded, clearly visible and bears long, eversible proboscis which lacks jaws; 2 conspicuous eyes; 5 antennae; 4 pairs of tentacular cirri.
**Body** 60–200 segments; parapodia with large, green, triangular dorsal cirri and small, ovoid ventral cirri.
**Colour** grass-green, sometimes with blacker or bluer shades.
**Habitat** in rock crevices in shallow water and on the lower shore where it sometimes creeps about over the rocks when the tide is out.
**Distribution** Mediterranean, Atlantic, English Channel, North Sea and west Baltic. N.B. the gelatinous, green egg cases of this worm may be found on the lower shore on sand or rocks, usually attached to seaweeds or pebbles.

egg mass of
*E. viridis*

*Eulalia
viridis*

### *Eulalia sanguinea* Oersted
(Not illustrated)
**Length** up to 60mm.
**Body** 60–140 segments; parapodia with leaf-shaped, grey-green dorsal cirri.
**Colour** very variable; white-pale green or brown, sometimes with paler dorsal line.
**Habitat** among seaweeds and stones from lower shore downward.
**Distribution** Mediterranean, Atlantic, English Channel and North Sea.

### Family Tomopteridae
Free-living, planktonic polychaetes whose bodies are transparent and bear large parapodia which lack chaetae and acicula.

### *Tomopteris helgolandica* (Greeff)
**Length** up to 17mm.
**Head** bears 2 conspicuous pallas and eyes; behind this are a pair of very short chaeta-bearing appendages, followed by another pair whose length is about two-thirds that of the body.
**Body** bilobed parapodia lack chaetae.
**Colour** transparent and colourless.
**Habitat** planktonic.
**Distribution** western Mediterranean, Atlantic and English Channel. N.B. various related species occur in all areas except the Baltic.
See Newell, G. E. and Newell, R. C. 1973 for more details of planktonic polychaetes.

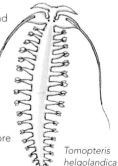

*Tomopteris
helgolandica*

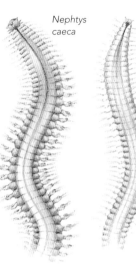

*Nephtys caeca*

## Family Nephtyidae

Medium to large, free-living polychaetes with flattened bodies somewhat rectangular in cross section. A small head bears a large eversible proboscis, which carries papillae and 4 horny jaws (see fig. 25, page 115), and 4 short antennae. Movement is by a characteristic rapid wriggling of the body.

### *Nephtys caeca* (O. F. Müller)

**Length** up to 250mm.

**Head** eyes not visible; form of head shown in fig. 24, page 115.

**Body** 90–150 segments; parapodia distinctly bilobed; dorsal cirrus very small, ventral cirrus conical; from the 4th chaeta bearing segment back, the parapodia bear a gill situated between the two groups of chaetae; characteristic thread at tail.

**Colour** various body colours (2 forms are shown); generally pearly grey with other shades.

**Habitat** burrowing in sand from middle shore down.

**Distribution** Atlantic, English Channel, North Sea and west Baltic.

## Family Glyceridae

Small to medium free-living polychaetes whose bodies taper at both ends. An eversible proboscis carries papillae and several, often 4, jaws. Antennae are minute.

### *Glycera convoluta* Keferstein

*Glycera convoluta*

**Length** up to 100mm.

**Head** anterior end reduced to small prostomial process with 4 minute antennae, bulbous, eversible proboscis with 4 jaws and many fine cylindrical papillae; form of head shown in fig. 25, page 115.

**Body** round and transparent; 120–180 segments each marked by 2 annular rings.

**Colour** tinted red-pink.

**Habitat** in sand or mud, often among seaweeds.

**Distribution** Mediterranean, Atlantic, English Channel and North Sea. These worms can bite an incautious collector.

## Family Eunicidae

Large, free-living polychaetes. The head has 2 eyes, 5 antennae, 2 bilobed palps and 1 pair of tentacular cirri (see fig. 26, page 115). The first two apparent body segments lack parapodia and chaetae. Large gills are present.

### *Eunice harassii* Audouin & Milne-Edwards

*Eunice harassii*

**Length** up to 250mm.

**Head** form shown in fig. 26, page 115.

**Body** parapodia with comb-like gills from about the 4th segment back.
**Habitat** lower shore downward under stones and rocks.
**Distribution** Mediterranean, Atlantic and English Channel.

### *Marphysa bellii* (Audouin & Milne-Edwards)
**Length** up to 200mm
**Head** eyes small, antennae weakly annulated.
**Body** long and filiform; 200–300 segments with chaetae; comb-like gills between about segments 12–35.
**Colour** body pink; gills red.
**Habitat** lower shore in sand and mud, and in shallow water.
**Distribution** Mediterranean, Atlantic and English Channel.

*Marphysa bellii*

### *Marphysa sanguinea* (Montagu)
Similar to *Eunice harassii* (above).
**Length** 300 to 600mm.
**Head** antennae very short and not conspicuous.
**Body** flattened about 300 segments bear chaetae. gills not comb-like, but arise as a bunch of about 7 filaments.
**Habitat** in rock crevices and sometimes among seaweeds from lower shore downward.
**Distribution** Mediterranean, Atlantic and English Channel.

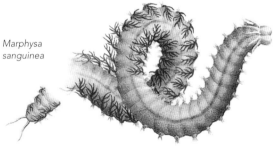

*Marphysa sanguinea*

*Ophryotrocha puerilis*

### *Ophryotrocha puerilis* Claparède & Mecznikow
**Length** up to 10mm.
**Head** bears 2 eyes, 2 small antennae and 2 small palps.
**Body** quite short and stubby, tapering slightly towards the extremities, 20–30 segments; parapodia bilobed without gills and with chaetae arranged in 2 groups.
**Colour** whitish and transparent, so that the jaws can be seen within.
**Habitat** among other invertebrates, e.g. bryozoans, ascidians and echinoderms.
**Distribution** Mediterranean, Atlantic, English Channel and North Sea.
N.B. some of the largest polychaetes of the Mediterranean belong to this family, e.g. *Eunice gigantea* Chiaje, 3m; *Halla parthenopeia* (Chiaje), 800mm; and *Diopatra neapolitana* Chiaje, 500mm.

125

### Family Orbiniidae

Polychaetes which usually burrow in mud or sand. The head generally lacks appendages but possesses 2 eyes. The body bears numerous segments and is divided into two regions, of which the thoracic half is flattened and enlarged, and the abdominal half is long and half cylindrical. Parapodia bear simple gills on the dorsal surface.

*Orbinia latreilli*

***Orbinia latreilli*** (Audouin & Milne-Edwards)
**Length** up to 400mm.
**Head** lacks eyes and appendages.
**Body** flat above, rounded below; very long and fragile; 400 segments bear chaetae, of which 35 may be in the thoracic region; gills borne on 5th chaeta-bearing segment back two terminal threads on tail; form of 26th parapodium shown in fig. 5, page 114.
**Colour** generally pink anteriorly and yellowish posteriorly.
**Habitat** on seabed in sand and mud.
**Distribution** Atlantic, English Channel and North Sea.

***Scoloplos armiger*** (O. F. Müller)
**Length** up to 150mm.
**Head** no appendages are visible, but 2 eyes may be seen sunk deeply in the head.
**Body** long; up to 200 segments bearing chaetae, of which up to 20 may be in the flattened thoracic region; gills are generally borne anywhere from the 9th chaeta-bearing segment backward; form of abdominal parapodia is shown in fig. 6, page 114.
**Colour** generally pink or orange.
**Habitat** on the seabed in sand or mud and among *Zostera* (see pages 64–65).
**Distribution** Atlantic, English Channel, North Sea and the west Baltic.

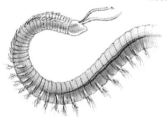

*Scoloplos armiger*

### Family Spionidae

Polychaetes which usually burrow in sand or in mud. The head is often characterized by 2 frontal horns, 4 eyes and 2 palps. The dorsal and ventral parapodial; cirri are often lamella-like, gills are carried dorsally on a number of parapodia. The body is not apparently regionalized.

### *Malacoceros fuliginosa* (Claparède)
(Not illustrated)
**Length** up to 60mm.
**Head** bears 2 frontal horns, 4 eyes and 2 banded palps.
**Body** long and slender; about 150 segments; gills borne on the first chaeta-bearing segment and backwards from it; form of anterior parapodia is shown in fig. 8, page 114.
**Colour** variable; often reddish.
**Habitat** on the seabed burrowing in tubes in sand and mud; sometimes under pebbles; occasionally in colonies.
**Distribution** Mediterranean, Atlantic, English Channel and North Sea.

*Scolelepis squamata*

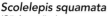

### *Scolelepis squamata*
(Söderström)
**Length** up to 80mm.
**Head** bears 4 small eyes disposed in a square, and 2 long palps which can reach back as far as the 24th chaeta-bearing segment when straightened out.
**Body** long, slender; about 200 chaeta-bearing segments; gills developed from about the 2nd segment back, but a few may be absent around the 10th segment; form of 45th parapodium shown in fig. 7, page 114.
**Colour** variable, but mainly blue-green; male whiter and female very green.
**Habitat** on the seabed in sand or mud.
**Distribution** Mediterranean, Atlantic, English Channel and North Sea.

### *Polydora ciliata* (Johnston)
**Length** up to 30mm.
**Head** bears 4 eyes arranged in a square and 2 long, thin palps.
**Body** relatively thin; up to 180 chaeta-bearing segments, 5th segment bears conspicuous chaetae; parapodia from the 7th as far back as the 10th before the last bear gills; fan-shaped tail appendage; form of 1st parapodium is shown in fig. 9, page 114.
**Colour** variable; usually brown-yellow.
**Habitat** found boring into oysters, *P. ciliata* may be considered a pest in oyster beds.
**Distribution** Mediterranean, Atlantic, English Channel, North Sea and west Baltic. N.B. only the palps may at first be visible as minute, fine threads protruding from the holes in the shell.

*Polydora ciliata*

## Family Magelonidae

Burrowing polychaetes, with an oval, flattened 1st segment. A large proboscis is present, but no antennae nor eyes. They possess 2 long palps with papillae. The body is divided into two regions; gills are lacking and parapodia have 2 lobes. Lamellae-shaped dorsal and ventral cirri. One genus is known.

*Magelona mirabilis*

### *Magelona mirabilis* (Johnston)

**Length** up to 170mm.
**Head** typical proboscis large.
**Body** 150 segments bear chaetae, of which the first 8 differ from the remainder; remaining parapodia have large dorsal, and small ventral, lamellae; form of 8th parapodium shown in fig. 10, page 114.
**Colour** palps and anterior end pale pink; posterior grey or greenish with white on sides.
**Habitat** in sand in shallow and deeper water.
**Distribution** Mediterranean, Atlantic, English Channel and North Sea.

*Chaetopterus variopedatus*

## Family Chaetopteridae

Tube-dwelling polychaetes with soft bodies divided into three regions. No eversible proboscis, but 2 eyes and 2 or 4 palps are present on the head. The anterior region of the body has few segments; the middle section has very few, highly modified segments bearing conspicuous, bilobed parapodia, posterior segments numerous.

### *Chaetopterus variopedatus* (Renier)

**Length** up to 250mm.
**Head** broad, bears 2 palps and a large, terminal mouth.
**Body** first 9 or so body segments bear chaetae; middle region characterized by a 'waist', and segments bearing 3 flap-like paddles ventrally; posterior part has bristled parapodia.
**Habitat** in parchment-like, u-shaped tubes up to 400mm long; often buried in mud or sand in shallow and deeper water.
**Distribution** Mediterranean, Atlantic, English Channel and North Sea.

## Family Cirratulidae

Sedentary polychaetes with cylindrical, tapering bodies. The head sometimes has eyes but lacks appendages. One anterior segment bears either distinct palps, or several tentacular filaments. The parapodia are bilobed, lack dorsal and ventral cirri, and bear simple, long, contractile gill filaments which show 2 conspicuous blood vessels (see fig. 11, page 114).

### Cirriformia tentaculata (Montagu)
**Length** up to 200mm.
**Head** indistinct; no eyes, but several pigment spots.
**Body** swollen dorsally and flattened or concave ventrally; about 300 compressed segments; gills on all chaeta-bearing segments except the last few.
**Habitat** in sand, mud and under pebbles from lower shore downward; often only the blood-red tentacles apparent.
**Distribution** Atlantic, English Channel and North Sea.

### Cirratulus cirratus (O. F. Müller)
**Length** up to 120mm.
**Head** indistinct; bears 4–8 eyes arranged in rows on both sides of 1st segment.
**Body** up to 130 segments; gills carried on all chaeta-bearing segments: form of parapodia shown in fig. 11, page 114.
**Habitat** and **Distribution** similar to *Cirriformia tentaculata*, but sometimes found in cracks of rocks.

## Family Flabelligeridae
Burrowing polychaetes. The chaetae of anterior segments surround the head which has a retractile mouth siphon, eyes, 2 large palps and retractable gills. Parapodia are bilobed, and the blood is green.

### Flabelligera affinis M. Sars
**Length** up to 60mm.
**Body** up to 50 segments.
**Habitat** under pebbles and on seaweeds in mud, and on sea-urchins.
**Distribution** Atlantic, English Channel, North Sea and west Baltic.

### Stylarioides plumosa
(O. F. Müller)
**Length** up to 60mm.
**Head** bears 4 eyes, 2 fat palps and 8 cylindrical gills.
**Body** 60–70 segments bear chaetae.
**Habitat** in mud and in cracks of rock from the lower shore downward.
**Distribution** Atlantic, North Sea and English Channel.

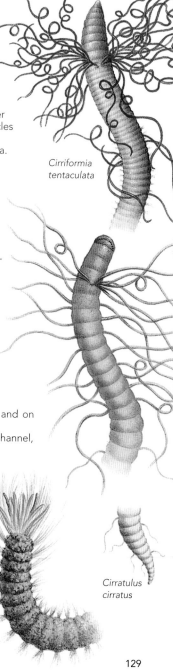

*Cirriformia tentaculata*

*Cirratulus cirratus*

*Flabelligera affinis*

*Stylarioides plumosa*

129

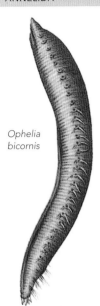

*Ophelia
bicornis*

## Family Opheliidae

Polychaetes with a conical preoral segment. The head has an unarmed proboscis; the main eyes are hidden by the skin, but lateral eyes are sometimes visible. Other head appendages are lacking. The body is short, convex dorsally and concave ventrally for all or part of its length. Bilobed parapodia are generally reduced and lack dorsal cirri, but ventral cirri are occasionally present.

### *Ophelia bicornis* Savigny

**Length** up to 60mm.
**Head** form shown in fig. 12, page 114.
**Body** about 32 segments bear chaetae; conspicuous ventral gutter runs from the 10th chaeta-bearing segment to the tail. 15 pairs of gills.
**Habitat** on the seabed, often in loose sand.
**Distribution** Atlantic and English Channel.

## Family Capitellidae

Polychaetes with a conical, retractile preoral segment. The head has a large, unarmed proboscis, and the mouth opens ventrally; 2 eyes are present. The body is earthworm-like, divided into two, with the anterior part short and often slightly swollen, The posterior region is thinner, much longer, and bears small, often twisted, gills and bilobed parapodia.

### *Capitella capitata* (Fabricius)

**Length** up to 100mm.
**Head** bears 2 small eyes ventrally.
**Body** tapers towards both ends, very variable, and not a typical example of the family; fragile; 90 or more segments bear chaetae.
**Habitat** often in dirty sand or under pebbles on lower shore and in shallow water.
**Distribution** Mediterranean, Atlantic, English Channel, North Sea and west Baltic. N.B. this species name is often used to describe a number of similar species and new ones are often being discovered.

*Capitella
capitata*

## Family Arenicolidae

Body composed of two or three distinct regions with many short segments. The head bears an unarmed proboscis but lacks antennae and palps. The parapodia are bilobed, with a conical dorsal lobe and a twisted ventral lobe.

juvenile in
mucus sheath

### *Arenicola marina* (Linnaeus)
**Lugworm**
**Length** up to 200mm.
**Head** bears eversible proboscis covered by papillae.

*Arenicola
marina*

Body cylindrical; 6 swollen anterior segments without gills are followed by 13 segments with gills; gill filaments short and bushy-like, not connected by a basal membrane; posterior part of body less swollen.
Habitat burrowing in sand from middle shore downward.
Distribution Mediterranean (rarely), Atlantic, English Channel, North Sea and west Baltic. N.B. a second species *Arenicola defodiens* (not illustrated) Cadman and Nelson-Smith has long pinnate gill filaments, these are connected by a basal membrane.

*Arenicolides ecaudata*

### *Arenicolides ecaudata* (Johnston)
Length up to 250mm.
Head lacks eyes and appendages.
Body tapers slightly at both ends; 40–60 segments bear chaetae; gills on all segments from about the 15th back.
Habitat in twisting burrows in sand or mud from the lower shore downward.
Distribution Atlantic, English Channel and North Sea.

## Family Maldanidae
### Bamboo worms
Cylindrical bodies with a trunk at both ends. The body is not divided into distinct regions, and there are relatively few, long segments. Shape slightly bamboo-like.

### *Euclymene lumbricoides* (Quatrefages)
Length up to 150mm.
Head eyes rarely visible; form of head shown in fig. 15, p114.
Body thin, with 19 segments bearing chaetae, tapering slightly after the 15th segment; preoral segment is conical.
Habitat burrowing in sand from lower shore downward.
Distribution Mediterranean, Atlantic and English Channel.

*Euclymene lumbricoides*

anterior lateral view

detail of tail

*Maldane sarsi*

### *Maldane sarsi* Malmgren
Length up to 100mm.
Head oval, keel-shaped, of complex form, being convex and bordered by a membrane cut into, but not scalloped (as in *Euclymene lumbricoides*); the segment bearing the mouth and the next 3 are all similar and glandular; no eyes; form of head shown in fig. 13, page 114.
Body 19 segments bear chaetae which occur in three forms: bent, barbed and straight.
Habitat on muddy substrates often in deep water.
Distribution Atlantic and North Sea.

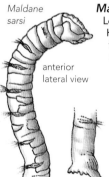

anterior lateral view

detail of tail

### Family Oweniidae

Tube-dwelling polychaetes with cylindrical bodies. They possess few segments, the anterior segments being longer than the posterior segments. The head lacks appendages and is capped by a small, folded membrane. All members of the family inhabit tubes attached generally to shells or stones.

Owenia
fusiformis

details of
tube tip

### *Owenia fusiformis* Delle Chiaje

**Length** up to 100mm.
**Head** typical and with 6 branched gills.
**Body** 20–30 segments; 3 short segments follow head, then 5-7 long segments after which the remainder are shorter.
**Colour** green-yellow.
**Habitat** in membranous tube with grains of sand or shell debris attached; generally on muddy sand from lower shore downward; part of the tube usually apparent.
**Distribution** Mediterranean, Atlantic, English Channel and North Sea.

tube

### Family Sternaspididae

Polychaetes with short bodies and short segments. The anterior chaetae are short and robust. Posterior segments bear filamentous gills and long chaetae.

### *Sternaspis scutata* (Ranzani)

Typical of family
**Length** up to 30mm.
**Head** reduced, lacks appendages, mouth ventral.
**Body** 20–22 body segments; first 3 segments bear chaetae arranged in arcs each side; 2 conspicuous genital papillae on 7th segment; posterior segments have long gills, conspicuous shield.
**Colour** white-grey-yellow.
**Habitat** in sand and mud.
**Distribution** Mediterranean, Atlantic, and English Channel.

*Sabellaria
alveolata*

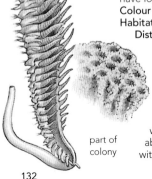

*Sternaspis
scutata*

### Family Sabellariidae

Tube-dwelling polychaetes with cylindrical bodies divided into three parts. The head end has modified chaetae set in up to 3 concentric rings to form a stopper for the tube; this is followed by 2 segments with reduced chaetae plus 3–4 other segments; the abdominal region is composed of about 30 segments with parapodia. A slender terminal region is present.

part of
colony

### Sabellaria alveolata (Linnaeus)
Typical of family
**Length** up to 40mm.
**Body** has 32–37 segments.
**Habitat** in tubes arranged in colonies and made of large sand grains; encrusting rocks and shells from lower shore downward.
**Distribution** Mediterranean, Atlantic, English Channel and North Sea.

## Family Pectinariidae
Short, stumpy, tube-dwelling polychaetes whose bodies are divided into three parts. The thoracic part carries modified head with gills and includes the first three segments with chaetae; the abdomen has bilobed parapodia, the short tail region is concave dorsally.

### Pectinaria koreni (Malmgren) ( =Lagis koreni)
**Length** up to 50mm.
**Head** shielded dorsally by chaetae; bears club-like papillae.
**Body** typical of family.
**Habitat** in tubes made of medium-sized sand grains, lying in sand with worm upside down.
**Habitat** lower shore downward.
**Distribution** Mediterranean, Atlantic, English Channel, North Sea and west Baltic.

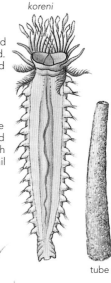

*Pectinaria koreni*

tube

## Family Terebellidae
Tube-dwelling polychaetes whose bodies are divided into two regions. The swollen thorax bears a reduced head with eyes and modified segments bearing many tentacles and blood-red branching gills. The abdomen tapers with reduced appendages. The tubes are membranous and covered with mud, sand, etc., and are buried or fixed to stones or plants.

*Amphitrite gracilis*

### Amphitrite gracilis (Grube)
**Length** up to 120mm.
**Body** long and gelatinous; 100–200 segments; 2 pairs of gills.
**Habitat** in twisted burrows in sand or mud from lower shore downward.
**Distribution** Mediterranean, Atlantic, English Channel and North Sea.

### Amphitrite johnstoni
Malmgren
**Length** up to 250mm.
**Body** has 90–100 segments; 3 pairs of gills.
**Habitat** and **Distribution** similar to *A. gracilis* (above).

*Amphitrite johnstoni*

133

### *Lanice conchilega* (Pallas)
#### Sand Mason
Typical of family.
**Length** up to 300mm.
**Head** mouth-bearing segment has 2 triangular lobes; short 2nd segment lacks appendages; 2 leaf-like lobes on 3rd segment, eyes may be visible; 3 pairs of gills.
**Body** 150–300 segments; swollen thoracic region consists of 7 chaeta-bearing segments; abdomen thin and fragile.
**Tube** composed of moderate to large sand grains with characteristic frayed appearance at the top, top projects from sand.
**Habitat** from middle shore downward.
**Distribution** Mediterranean, Atlantic, English Channel and North Sea.

*Lanice conchilega*
removed from tube

tube of *Lanice
conchilega*

## Family Salbellidae
Tube-dwelling polychaetes whose bodies are divided into two regions; often flattened, with a short thoracic part, and a long abdomen. The head is reduced and bears stiff feather-like tentacles forming a crown of flower-like gills borne on the 1st segment, so as to surround the mouth. The tube is mucoid, with particles of sand and mud usually embedded in it.

### *Sabella pavonina* Savigny
#### Peacock Worm
**Length** up to 250mm.
**Head** reduced; bears 1 pair of palps and 2 semicircular clusters of variably patterned gills each consisting of 8–45 filaments and which may superficially appear united to form a single crown.
**Body** rounded above and flattened below; 100–600 segments; thorax 6–12 segments bearing chaetae.
**Colour** variable.
**Habitat** in membranous tubes standing free from substrate; in mud and sand in shallow water.
**Distribution** Mediterranean, Atlantic, English Channel and North Sea.

*Sabella
pavonina*

### *Sabella spallanzani* (Gmelin)
#### ( =*Spirographis spallanzani*)
**Length** 200–300mm.
**Head** reduced, bearing 2 inequal groups one larger than the other, of many gill filaments (up to 300) twisted into a spiral crown; no eyes.
**Body** 100–300 chaeta-bearing segments of which 8 are thoracic; cyclindrical in shape, tapering suddenly at the posterior.
**Tube** cylindrical and encrusted, often rubbery.
**Colour** yellow, maroon or brown; gills of variable colour: white, violet, yellow or brown, sometimes patterned.

### *Diodora graeca* (Linnaeus)
**Keyhole Limpet**
**Shell** up to 40mm long, conical and ribbed with charac-
teristic keyhole at apex (when alive a small siphon
protrudes here); no trace of coiling, animal's mantle
may expand around the base of the shell.
**Colour** greyish.
**Habitat** on rocks on lower shore and down to 20m.
**Distribution** Atlantic, English Channel and North Sea.

*Diodora
graeca*

### *Tectura virginea* (O. F. Müller)
**White Tortoiseshell Limpet**
**Shell** up to 12.5mm long; flattened, smooth, delicate cone
shape with the apex offset towards the front of the shell.
**Colour** white-pinkish.
**Habitat** on lower shore and down to 10m, especially associated
with *Laminaria* (see pages 32–33).
**Distribution** Mediterranean, Atlantic, English Channel, North
Sea and west Baltic.

*Tectura
virginea*

### *Tectura tessulata* (O. F. Müller)
**Tortoiseshell Limpet**
Similar to *T. virginea* (above), but markings are reminiscent
of tortoiseshell.
**Shell** up to 25mm long.
**Habitat** on rocks, etc., on lower shore and in shallow water.
**Distribution** Atlantic, North Sea and west Baltic.

*Tectura
tessulata*

**Note on the genus Patella** A number of species of this genus
occur in the European area, of which four are treated here. They
are all characterized by a conical, rough, ribbed shell whose
apex lies towards the front of the animal rather than centrally.
There is no opening in the apex as there is in the keyhole
limpets. The shell fits tightly and exactly against the substratum.
A careful examination of the mantle tentacles (which may be
visible protruding from the periphery of the body under the lip
of the shell when the animal crawls in a dish of sea water) and
of the interior of the shell, may help distinguish the species.
For really accurate identification it may be necessary to
dissect a specimen and examine the inside of the shell
and the arrangement of the teeth on the radula (see
Fretter & Graham. 1962. pp 495 and 496). All four
species of *Patella* illustrated have been painted with
the front of the shell uppermost. Only the interiors are
shown for *Patella* but for *Helcion* the exterior is shown.
Several species of limpet often occur on the same
rocky shore at different levels.

### *Patella vulgata* Linnaeus
**Common limpet**
**Shell** up to 70mm long; tall, often with a blunt tip; ribs irregular.
**Colour** exterior greenish blue or grey, often encrusted with
barnacles (see pages 202–205); interior white or yellow, with
a white-brown scar left by the animal inside the shell apex;

old shell interior
of *Patella vulgata*

*Patella ulyssiponensis*

marginal mantle tentacles transparent. N.B. illustration shows interior of an old shell.

**Habitat** on rocks on upper and middle shore; the shells becoming less tall the further down they occur, and individuals less numerous where seaweeds are thick; often found in exposed places.

**Distribution** Atlantic, English Channel and North Sea. N.B. the shell illustrated is an old one.

### *Patella ulyssiponensis* (Gmelin) *( =P. aspera)*

**Shell** up to 70mm long; quite flattened; exterior lacking contrasting rays.

**Colour** interior white, sometimes with blue iridescence; foot orange; marginal mantle tentacles cream.

**Habitat** on rocks from about the centre of the middle shore down to the bottom of the lower shore; generally found in exposed places.

**Distribution** Atlantic north to south-west Britain, and North Sea (Scotland to Norway).

*Patella depressa*

### *Patella depressa* Pennant *( =P. intermedia)*

**Shell** up to 40mm long, less tall than *P. vulgata (above)*, ribs finer; exterior and interior margins may show dark rays.

**Colour** interior with orange-cream scar; foot dark; marginal mantle tentacles opaque.

**Habitat** on rocks on the middle shore often found in exposed places.

**Distribution** Atlantic north to Anglesey (North Wales), English Channel east to Isle of Wight; generally absent from Ireland.

### *Patella lusitanica* (Gmelin)

**Shell** up to 40mm long; less long, narrower, but taller than other species.

**Colour** outer surface of shell spotted black; interior with dark rays running a little way in; pale scar.

**Habitat** on rocks, usually on relatively vertical surfaces on upper half of the middle shore.

**Distribution** Mediterranean and Atlantic north to Biscay.

*Patella lusitanica*

*Helcion pellucidum*

### *Helcion pellucidum* (Linnaeus)
**Blue-rayed Limpet**
**Shell** about 15mm long; smooth and semi-transparent.

**Colour** rows of beautiful, bright blue spots running from the top to the margin, may fade with age.

**Habitat** generally attached to the fronds and holdfasts of *Laminaria* (see page 33) on the lower shore and in shallow water.

**Distribution** Atlantic, English Channel and North Sea. N.B. younger animals sometimes occur on the fronds and may appear brighter than older individuals located more often on the holdfasts.

young specimen          older specimen

Note on the genera Gibbula and Monodonta These topshells are conical in shape and circular in section. The inner layer is mother-of-pearl and may show through if the shell is worn. A horny or calcareous operculum closes the mouth of the shell when the snail's foot is withdrawn.

### *Gibbula magus* (Linnaeus)
**Turban Topshell**
Shell about 20mm wide but less high; about 5 whorls (sometimes more); upper surface of the whorls is bumpy, suture between whorls pronounced; umbilicus conspicuous.
Colour yellow-white with red or purple marks.
Habitat buried in sand down to about 10m.
Distribution Mediterranean, Atlantic and English Channel.

*Gibbula magus*

### *Gibbula umbilicalis* (da Costa)
**Purple Topshell**
Shell about 12.5mm high; rather wider than high; generally like a compressed cone with slightly convex outline; there may be 7 whorls, no noticeable steps between whorls; umbilicus conspicuous.
Colour green-grey with conspicuous purple stripes.
Habitat on rocks on the middle shore and the upper part of the lower shore.
Distribution Atlantic and English Channel.

*Gibbula umbilicalis*

### *Gibbula pennanti* (Philippi)
**Pennant's Topshell**
(not illustrated) similar to *G. umbilicalis* (above) but with profile of shell more 'stepped' between whorls. Umbilicus open in young shells but closing as they age. Up to 15mm high.
Colour purple banding not as regular as in *G. umbilicalis*.
Habitat on rocky shores from the middle shore downwards, sometimes in pools.
Distribution Mediterranean north to Channel Islands.

*Gibbula cineraria*

*Monodonta lineata*

### *Gibbula cineraria* (Linnaeus)
**Grey Topshell**
Shell about 12.5mm high and of similar width; up to 7 whorls which may appear somewhat compressed; umbilicus small.
Colour greyish shell with darker grey-red markings in the form of very narrow bands which may be somewhat faded.
Habitat under stones and on seaweeds on the lower shore and down to about 20m.
Distribution Atlantic, English Channel and North Sea.

tooth —

### *Monodonta lineata* (da Costa) *( =Gibbula lineata)*
**Toothed Winkle** or **Thick Topshell**
Shell about 25mm high and of similar width; quite conical with about 6 poorly defined whorls and slight umbilicus (in some specimens this is obscured by growth); mouth shows mother-of-pearl which usually stretches to the umbilical region.

Colour grey-green with purple zig-zags; top of shell may be worn away and appear pearly yellow.
Habitat on rocks on the middle shore.
Distribution Atlantic north to Anglesey and west English Channel, N.B. 'tooth' on the inside of the mouth opening easily distinguishes this shell from *Littorina littorea* (see page 154).

### *Jujubinus striatus* (Linnaeus)
### Grooved Topshell

*Jujubinus striatus*

Shell about 10mm high, but less wide; steep cone shape with almost straight sides; bottom whorl has about 6 spiral ridges of which the basal one is most developed and gives a keel effect; umbilicus lacking.
Colour grey-white with brown-red vertical markings.
Habitat on soft substrates from extreme lower shore down to about 100m.
Distribution Mediterranean, Atlantic north to the south-west of the British Isles and west English Channel.

### *Calliostoma zizyphinum* (Linnaeus)
### Common Topshell or Painted Topshell

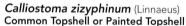

Shell about 25mm high and of similar width; strongly conical with straight sides; about 9 whorls which are relatively shallow; umbilicus lacking.
Colour yellow-pink mottling with brown or red stripes.
Habitat on rocks and under stones on lower shore and down to 100m.
Distribution Mediterranean, Atlantic, English Channel and North Sea.

*Calliostoma zizyphinum*

### *Astraea rugosa* (Linnaeus)
### Rough Star-shell

Shell about 50mm high and of similar width; thick and heavy with about 7 whorls which are moulded into bumps on their upper surface and have thorny, spiral lines on their sides; foot of snail carries a calcareous operculum with spiral markings (here shown stopping the aperture).

*Astraea rugosa*

Colour usually red-brown.
Habitat on rocks from the lower shore downward.
Distribution Mediterranean and Atlantic coasts of Spain and Portugal.

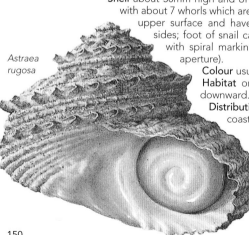

### *Tricolia pullus* (Linnaeus)
**Pheasant Shell**
**Shell** about 8mm high but less wide; generally about 4 whorls, of which the last takes up more than half the total height; foot of snail bears conspicuous white calcareous operculum.
**Colour** white and glossy with irregular brown-red markings.
**Habitat** in rock pools on the extreme lower shore, especially among red seaweeds; sometimes found on sandy substrates.
**Distribution** Mediterranean, Atlantic and English Channel.

*Tricolia pullus*

### *Theodoxus fluviatilis* (Linnaeus)
**Shell** about 10mm wide, but less high and of unusual shape; being flattened; mouth partly blocked by columellar growth; outer lip usually thickened; foot of snail bears calcareous operculum.
**Colour** speckled.
**Habitat** freshwater and estuarine regions where the salinity does not exceed 0.6% and where the calcium content is high; in situations where the animal may be protected from water currents and strong light, e.g. under stones and wood.
**Distribution** may be common in areas bordering the Atlantic region where the right conditions prevail; in the Baltic it is found on brown seaweeds and stones, occurring with increasing frequency in more open, turbulent water, providing the salinity is low.

*Theodoxus fluviatilis*

## Order Mesogastropoda
### Periwinkles, tower shells, worm shells, etc.
Algae eaters, deposit feeders and predators. The inner shell lacks mother-of pearl but often has a horny operculum.

**Note on the chink shells, family Lacunidae** Four species of chink shell may be found in north-west Europe, usually on the lower shore, often associating with algae, or in shallow water. The shells are normally small, delicate and smooth, and the last whorl is big in relation to the rest of the spire. The term 'chink' refers to the groove or 'chink' on the columella which characterises these shells.

### *Lacuna vincta* (Montagu)
**Banded Chink Shell**
**Shell** about 80mm high, sometimes more, but less wide; generally of conical shape with about 5 smooth whorls; apex pointed, mouth oval; small but deep umbilicus.
**Colour** green-yellow and semi-transparent with red-brown banding.
**Habitat** lower shore and in shallow water on seaweeds (e.g. *Ceramium* and *Polysiphonia*, see pages 54 and 60–61).
**Distribution** Atlantic, English Channel, North Sea and west Baltic.

*Lacuna vincta*

**Note on the genus Littorina** The following five species are all characterized by solid shells lacking an umbilicus, but possessing smooth columellas. The foot of the snail carries a horny operculum. These species are generally common on rocky shores where they have each evolved to fit a particular niche. They can variously withstand exposure to air, desiccation and lowered salinity and their respiratory and reproductive processes show adaptations to the particular position they have taken up on the shore. Heller, J. 1975 revised the taxonomy of some British littorinids. See also Reid, D. G. 1996.

### *Littorina obtusata* (Linnaeus) *( =L. littoralis)*
**Flat Periwinkle**
**Shell** up to 10mm high, occasionally larger; spire very compressed so that the mouth and the last whorl comprise virtually all the height; shell apparently smooth but if examined with a hand lens, very fine sculpturing can be seen.

**Colour** variable; brown, red, green, orange and yellow, may be banded.

**Habitat** on seaweeds (especially *Fucus vesiculosus* and *Ascophyllum nodosum* see pages 38–39) on the lower middle, and upper parts of the lower shores.

**Distribution** Atlantic, English Channel, North Sea and west Baltic. N.B. this winkle is a gill breather and feeds on seaweeds.

*Littorina obtusata*

### *Littorina fabalis* (Turton) *( =L. mariae)*
(Not illustrated) Similar to *L. obtusata* above but shell so compressed as to completely lack spire. Outer lip meets shell coils high up.
**Habitat** usually lower on shore than *L. obtusata* otherwise similar to it.
**Distribution** as *L. obtusata*.

**Note on the Rough Periwinkles** All these periwinkles usually occur on upper shore and the upper part of the middle shore. They vary in size from 4.5mm to 10mm in height, as well as in colour and shell ornamentation. Distinguishing between these species is a specialist's task and one in which not all authorities agree. It may be easiest for the non-specialist student to regard them as members of a species aggregation, *Littorina saxatilis*.

### *Littorina saxatilis* (Olivi)
**Rough Periwinkle**
The status of this species has been the subject of much discussion (See Reid, 1996). It is ovoviviparous (produces fully

formed eggs that are retained inside the female body until live offspring are released). Conse-quently there are many local varieties as no larvae are distributed via the plankton and different forms may be found over a small distance of the shore. Several varieties have been recognised e.g. *L. tenebrosa* and *L. neglecta* (See Hayward and Ryland 1998).
**Shell** about 8mm high; 6–9 whorls separated by conspicuous sutures; each whorl has ridges and grooves which make the shell rough to the touch. Outer lip of the opening meets the spire almot at right angles.
**Colour** variable; red-black.
**Habitat** on upper shore and upper middle shore often in crevices; generally more plentiful where the beach is exposed.
**Distribution** Atlantic, English Channel and North Sea and west Baltic.

*Littorina saxatilis*

### *Littorina arcana* Hannaford Ellis
(Not illustrated) Similar to *L. saxatilis* (above) but lays eggs in masses from which young snails hatch. No larval stage in the plankton. This distinction is not a 'field character' direction needed to distinguish from other species.
**Habitat** upper middle shore in exposed places.

### *Littorina compressa* Jeffreys *( =L. nigrolineata)*
(Not illustrated) Similar to *L. saxatilis* (above) but usually with dark brown pigmentation between the wide, flat-topped, strap-like ridges of the coiled shell.

### *Melaraphe neritoides*
(Linnaeus) *( =Littorina neritoides)*
**Small Periwinkle**
**Shell** about 5mm high; sharply conical with point-ed apex; smooth surface and fragile appearance; outer lip of the opening roughly parallel to the spire where the two meet.
**Colour** blue-black.

*Melaraphe neritoides*

**Habitat** on extreme upper shore, usually in crevices; generally more plentiful where the beach is exposed. Has a sea going larva, so post larval young may be found returning from the sea at all levels of the shore.
**Distribution** Mediterranean, Atlantic, English Channel and North Sea. N.B. lung breather which feeds on lichens. Do not confuse it with small examples of *L. saxatilis* (above) which occur lower on the shore and are rough to the touch.

153

*Littorina
littorea*

### Littorina littorea (Linnaeus)
**Edible Periwinkle**
**Shell** about 25mm high; sharply conical with pointed apex and surface sculpturing; outer lip of opening is more or less parallel to the spire where the two meet.
**Colour** grey-black-brown-red and always patterned with concentric darker lines; columella white.
**Habitat** on rocks, stones and seaweeds on the middle and lower shores.
**Distribution** Mediterranean, Atlantic, English Channel, North Sea and west Baltic. N.B. a gill breather which feeds on seaweed. It may migrate up or down the shore during the breeding seasons. Do not confuse small specimens with *L. saxatilis* (see page 152) where the outer lip of the shell opening meets the spire at right angles, or with *Monodonta lineata* (see page 149). The black banded tentacles of *L. littorea* may also be seen.

*Truncatella
subcylindrica*

### Truncatella subcylindrica (Linnaeus)
**Looping Snail**
**Shell** about 5mm high; conspicuous because of the lack of apical whorls on the adult shell (these break off from the young shell when it reaches maturity); adult has about 3 whorls remaining; mouth ear-shaped.
**Colour** yellow-brown, with the shell surface finely ribbed.
**Habitat** in muddy places on the upper shore usually associated with seaweeds or stones.
**Distribution** Mediterranean and Atlantic. N.B. it moves by looping like a leech.

*Hydrobia
ulvae*

### Hydrobia ulvae (Pennant)
**Laver Spire Shell**
**Shell** about 6mm high and complete (possessing all apical whorls) as adult; conical shape terminates in blunt apex; whorls do not appear swollen; outer lip of opening is more or less straight edged where it joins the spire at an acute angle.
**Colour** brown to yellow; diamond-shaped area of pigment on the head between the eyes.
**Habitat** on mud in estuaries, normally on the middle shore and in other places where the water is brackish.
**Distribution** Atlantic, English Channel, North Sea and Baltic. N.B. the left tentacle is thicker than the right tentacle.

*Hydrobia
ventrosa*

### Hydrobia ventrosa (Montagu)
**Shell** about 6mm high and complete (possessing all apical whorls) as adult; conical shape often terminates in a sharper point than *H. ulvae* (above); whorls appear swollen; outer lip opening is curved and joins the spire at a right angle.
**Colour** brownish; v-shaped area of pigment on the head between the eyes.

Habitat often in lagoons without direct communication with sea water and where the water is brackish; in the Baltic it may be found associated with seaweeds and also on gravel substrates where there is some wave action.
Distribution Atlantic, English Channel, North Sea and Baltic.

### Potamopyrgus jenkinsi (Smith)
Shell about 5mm high; last whorl takes up about two-thirds total shell height and may be keeled or bear bristles.
Colour yellowish, but often blackened by deposits.
Habitat in muddy conditions among stones and seaweeds where the water is brackish and running; sometimes in lagoons on sand.
Distribution Atlantic, English Channel, North Sea and Baltic.
    Graham, A. 1988 provides more information on the identification of Hydrobia and Paludestrina species.

Potamopyrgus
jenkinsi

### Bithynia tentaculata (Linnaeus)
Shell about 10mm high; conical and smooth; umbilicus minute; about 6 whorls with conspicuous growth lines showing the previous positions of the mouth; mouth wide.
Colour brown, with dark growth lines.
Habitat in fresh water, on vegetation in Summer and on mud in Winter.
Distribution Baltic; may occur elsewhere in ponds, canals and rivers.

Bithynia
tentaculata

### Alvania cancellata (da Costa)
Shell about 3mm high; spiral and longitudinal lines mark the surface.
Colour brown-pink-grey.
Habitat on rocks and gravel and among other organisms on the extreme lower shore and in shallow water.
Distribution Atlantic, English Channel and North Sea, related species occur in the Mediterranean and in the foregoing areas.

Alvania
cancellata

### Rissoa parva (da Costa)
Shell about 7mm high; slight ribs; the lips of the aperture are somewhat turned out. Two forms are recognised, the ribbed shell form shown here, and the smooth one.
Colour generally white-grey-brown, sometimes with a comma-shaped mark on the largest whorl.
Habitat extreme lower shore and shallow water, usually associated with coraline seaweeds and under stones; sometimes in pools.
Distribution Atlantic, west English Channel, North Sea and west Baltic. N.B. many other related species and varieties are known from the European area. One such is Cingula cingillus (Montagu) (not illustrated) which is a minute, conical, banded shell occurring in crevices and under stones especially in silty places on the middle and lower shore and in shallow water.

Rissoa
parva

*Turritella communis*

Note on small marine gastropods About 20 species of small snail belonging to the families Hydrobidae, Rissoidae, Barleeidae, Cingulopsidae, Skeneopsidae, Omalogyriidae and Rissoellidae occur in north-west European waters. Their treatment is beyond the scope of this book. Reference should be made elsewhere, e.g. to Graham 1988.

### *Turritella communis* Risso
**Tower Shell**
**Shell** 40–60mm high; many conspicuous whorls; relatively narrow for its height; spiral ridges on the whorls; mouth relatively small.
**Colour** variable; red-brown-yellow-white.
**Habitat** lies partly buried in sand or mud, sometimes associated with other organisms, generally down to about 80m.
**Distribution** Mediterranean, Atlantic, English Channel and North Sea.

### *Vermetus gigas* Bivone
**Giant Worm Shell**
**Shell** up to 200mm long; irregularly coiled tube.
**Colour** grey-white.
**Habitat** on soft and hard substrates, stones and shells.
**Distribution** Mediterranean. N.B. several other related species are found in Europe.

*Vermetus gigas*

### *Bittium reticulatum* (da Costa)
**Needle Shell**
**Shell** up to 150mm high and relatively narrow, many finely latticed whorls with minute tubercles.
**Colour** brownish.
**Habitat** under stones and among rocks on the lower shore and in shallow water.
**Distribution** Atlantic, English Channel, North Sea and west Baltic.

*Bittium reticulatum*

### *Aporrhais pespelecani* (Linnaeus)
**Pelican's Foot Shell**
**Shell** about 35mm; many whorls with conspicuous tubercles; last whorl bears a flared outer lip which is ribbed and produced into about 4 points, although this may not be apparent in juvenile specimens; the fluted lip generally shields the animal's head.
**Colour** greyish.
**Habitat** burrows in mud, sand or gravel down to about 80m.
**Distribution** Mediterranean, Atlantic, English Channel and North Sea.

*Aporrhais pespelecani*

### *Capulus ungaricus* (Linnaeus)
**Bonnet Limpet** or **Hungarian Cap Shell**
Shell about 50mm wide; bonnet-shaped; apex turned
backwards and slightly coiled; periostracum provides a
fringe to the opening.
**Colour** white-yellow marked by spiral ridges;
interior white, periostracum brown.
**Habitat** attached to other shells,
usually bivalves, from which it
steals food by means of its
long proboscis; normally in
deep water around 100m.
**Distribution** Mediter-
ranean, Atlantic, English
Channel and North Sea.

*Capulus
ungaricus*

### *Crepidula fornicata* (Linnaeus)
**Slipper Limpet**
Shell up to 25mm wide; oval-shaped; apex shows some coiling;
outer surface shows growth lines.
**Colour** yellow, white, green-brown, occasionally with red
markings; underside usually white.
**Habitat** usually attached to others of the same species
and to bivalves, e.g. mussels and oysters in shallow water.
**Distribution** Atlantic, English Channel and North Sea.
N.B. this species normally lives in groups and forms a
chain of individuals. In such cases the animals at the
bottom of the chain are the oldest and are females.
Those at the top start life as males and as they age,
become females. This species is a serious pest in oyster
beds, and was introduced from America with imported oysters.
It has now spread from East Anglia along the east coast to
Scotland and around the south coast to Wales.

*Crepidula
fornicata*

### *Calyptraea chinensis* (Linnaeus)
**China Man's Hat**
Shell up to 15mm across; round and limpet-like,
opening circular.
**Colour** White-yellow.
**Habitat** in shallow water, on stones and shells
often in muddy areas.
**Distribution** Mediterranean, Atlantic and west
English Channel.

*Calyptraea
chinensis*

### *Velutina velutina* (C. F. Müller)
**Velvet Shell**
Shell up to 20mm long; about 3 whorls of which the last is the
largest and has a wide opening; covered by a velvet-like
periostracum and partly enclosed by the thick yellow mantle
of the snail in life.
**Colour** brownish.
**Habitat** among *Alcyonium* and sea-squirts (see pages 101,
272–275) from the lower shore down to about 50m.
**Distribution** Atlantic, English Channel and North Sea.

*Velutina
velutina*

157

*Trivia
monacha*

### *Trivia monacha* (da Costa) *( =Cypraea europaea)*
**European Cowrie**
**Shell** about 12mm when measured along the slit-like opening; spire very short; polished surface traversed by about 20 delicate ribs.
**Colour** pink-purple-brown above and pale below; characteristically bearing about 3 conspicuous, dark brown spots above; in life the shell may be partly covered by the variously coloured mantle folds of the animal.
**Habitat** among rocks and compound ascidians on which it feeds and deposits its eggs on the lower shore and in shallow water.
**Distribution** Mediterranean, Atlantic, English Channel and North Sea.

### *Trivia arctica (Montagu)*
(Not illustrated) Similar in all respects to *T. monacha* (above), but generally slightly smaller and lacking the conspicuous dark brown spots on the upper part of the shell.
**Habitat** rarely found between the tidemarks, usually down to 100m.
**Distribution** Atlantic, English Channel and North Sea.

*Erronea
pirum*

### *Erronea pirum* (Linnaeus)
**Pear Cowrie**
**Shell** up to 50mm when measured along slit-like opening; outer and inner lip toothed.
**Colour** red-brown above with irregular darker markings; light red below.
**Habitat** on hard substrates in deep water and associated with seaweeds.
**Distribution** Mediterranean and Atlantic north to Portugal.

*Polinices
polianus*

### *Polinices polianus* (Delle Chiaje)
**Alder's Necklace Shell**
**Shell** about 15mm high; globular and with a low spire; sutures between whorls not very distinct; last whorl expanded; umbilicus half occluded by growth of shell; ear-shaped operculum.
**Foot** of snail appears large and may be partly reflected over the shell when the animal is active.
**Colour** shiny and white-yellow with light, red-brown patterns.
**Habitat** burrowing in sand where it hunts for bivalve prey, on lower shore and down to about 70m.
**Distribution** Mediterranean, Atlantic, English Channel and North Sea.

### *Polinices catienus*
**Spotted Necklace Shell**
(Not illustrated) Similar to *P. polianus* above, but spire of shell higher and sutures more stepped and conspicuous.
**Habitat** and **Distribution** similar to *P. polianus* but absent from Norway.

### *Epitomium clathrus* (Linnaeus)
#### Common Wentletrap
Shell up to 40mm high, often less; a number of whorls bearing conspicuous diagonal stripes; mouth round; operculum horny.

Colour varies from colourless to brown-red.

Habitat generally in water down to 80m, but migrates to rocks on the shore near sand and mud to spawn.

Distribution Mediterranean, Atlantic, English Channel, North Sea and west Baltic. N.B. animal has long proboscis.

*Epitomium clathrus*

### *Dolium galea* (Linnaeus)
#### Giant Tun shell
Shell up to 150mm high; characteristic spiral ridges and a low spire.

Colour white-brown-yellow.

Habitat in deep water.

Distribution Mediterranean and Atlantic coasts of Portugal and Spain.

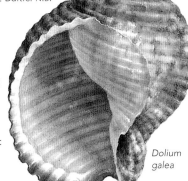

*Dolium galea*

### *Ianthina exigua* Lamarck
#### Violet Sea Snail
Shell about 15mm high; thin-walled and with about 5 whorls; delicate, v-shaped markings.

Colour violet (violet pigment may also be released).

Habitat a pelagic snail floating by means of a raft of trapped air bubbles and feeding on siphonophores; shells may occasionally be washed up after periods of westerly gales.

Distribution Atlantic.

In life

*Ianthina exigua*

### *Cassidaria echinophora* (Linnaeus)
#### Knobbed Helmet Shell
Shell up to 100mm across and higher than it is wide; bears large tubercles arranged in spirals; siphonal canal present; horny operculum.

Colour reddish-brown-grey.

Habitat on sandy substrates in both shallow and deep water.

Distribution Mediterranean and neighbouring Atlantic.

*Cassidaria echinophora*

# ORDER NEOGASTROPODA
## *WHELKS, ETC.*

Deposit feeders and predators. The inner shell lacks mother-of-pearl and an operculum is present. The snails have well developed siphons which are often supported by a siphonal groove or canal in the shell.

### *Trophonopsis muricatus* Montagu
**Shell** up to 1.9mm high. Whorls ornamented by ribs; long siphonal canal about the same length as the aperture height.
**Colour** white to yellowish sometimes with spiral brown banding.
**Habitat** from shallow water down on mud, sand and gravel.
**Distribution** Mediterranean and Atlantic.

### *Trophonopsis barvicensis* (Johnston)
(Not illustrated) Similar to *T. muricatus* (above) but with coarser ribs.
**Distribution** A northern species; North Atlantic and northern North Sea.

*Trophonopsis muricatus*

### *Nucella lapillus* (Linnaeus) *( =Thais lapillus)*
**Dogwhelk**
**Shell** about 30mm high with short siphonal canal; heavy with about 5 whorls which are marked with spiral lines; the last whorl being the largest; outer lip of the opening is thick and toothed in adults.

2 forms

**Colour** varies, but is basically ash-grey to cream, often marked or patterned with dark brown spirals; snail itself cream coloured.
**Habitat** on rocky shores, except those which are very exposed, in crevices and among barnacles (on which it preys) in the middle shore region.
**Distribution** Atlantic, English Channel and North Sea. N.B. egg capsules resembling grains of barley may be found (see page 314).

*Nucella lapillus*

### *Ocenebra erinacea* (Linnaeus)
**Sting Winkle** or **Oyster Drill**
**Shell** up to 60mm high, sometimes smaller; siphonal canal open in juvenile specimens, but closed in for the greater part of its length in older specimens so that it is tubular; unevenly sculptured; about 5 ribbed whorls of which the last is the largest; whorls have spiral lines or ridges; outer lip of opening toothed and thick.
**Colour** yellow-white with dark brown marks.
**Habitat** on muddy gravel, sand and rocks from the lower shore down to deep water (possibly to 100m), but migrating inshore to spawn.
**Distribution** Mediterranean, Atlantic, English Channel and North Sea. N.B. *O. erinacea* is a notable pest in oyster

*Ocenebra erinacea*

beds, attacking the oysters by means of the radula which functions like a drill, and then sucking out the oyster's flesh.

### *Tritonalia aciculata* (Lamarck)
**Shell** up to 20mm high; about 6 ribbed whorls; opening of shell toothed on the outer lip.
**Colour** brownish; interior of opening brown.
**Habitat** on rocks and sand from about 10m downward.
**Distribution** Mediterranean and neighbouring Atlantic coasts.

*Tritonalia aciculata*

### *Neptunea antiqua* (Linnaeus)
**Red Whelk**
**Shell** up to 100mm high; siphonal canal well developed, spire quite pointed, last whorl occupies abouth 80% of shell height, sutures quite clear, conspicuous ridge coils from the siphonal canal to the umbilical region.
**Colour** yellow to yellowish red.
**Habitat** shallow and deeper water on muddy bottoms.
**Distribution** Atlantic north from Biscay, Channel and North Sea.

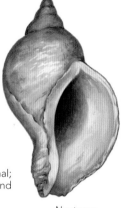

### *Buccinum undatum* Linnaeus
**Buckie** or **Common Whelk**
**Shell** up to 80mm high, sometimes more; short siphonal canal; spire pointed with well-defined whorls which are lined and ribbed; aperture large with smooth outer edge.
**Colour** pale brown.
**Habitat** on sand and mud from shallow water down to about 100m.
**Distribution** Atlantic, English Channel, North Sea and west Baltic. N.B. empty shells of this species are often inhabited by hermit crabs (see page 227). They are also sometimes covered with sponges, hydroids and anemones. The large, rounded, spongy egg masses of *Buccinum undatum* are often washed up on the shore, each hollow cell being an individual egg-case (see page 314).

*Neptunea antiqua*

### *Hinia reticulata* (Linnaeus)
**Netted Dogwhelk**
**Shell** up to 30mm high; short siphonal canal; conical; about 7 whorls which are not well defined and which are patterned in small squares due to the crossing of lines and ribs; oval aperture with thick outer lip ornamented by many teeth.
**Colour** brown.
**Habitat** under stones and in crevices, often in muddy areas on the lower shore and in shallow water.
**Distribution** Mediterranean, the Atlantic, English Channel, North Sea and west Baltic.

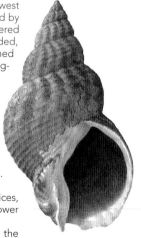

*Hinia reticulata*

*Buccinum undatum*

161

### Hinia incrassata (Ström)
**Thick-lipped Dogwhelk**
**Shell** up to 15mm high; conical; short siphonal canal; about 7 well-defined, convex whorls which are sculptured with lines and ribs; oval aperture with thick outer lip ornamented by seven teeth.
**Colour** brownish with dark bands.
**Habitat** under stones and in crevices, often in muddy places on the lower shore and in shallow water.
**Distribution** Mediterranean, Atlantic, English Channel and North Sea.

*Hinia*
*incrassata*

### Hinia pygmaea (Lamarck)
**Small Dog Whelk**
(Not illustrated) Similar to *H. incrassata* above but with pronounced 'netting' of the shell ribs. Up to 14mm high aperture wih 9–10 teeth on outer lip.

## SUBCLASS OPISTHOBRANCHIA

Gastropods which undergo torsion followed by de-torsion. The shell is reduced or absent. These animals often have conspicuous external gills and are brightly coloured. For identification of other opisthobranchs see Thompson, T. E. 1976, and Thompson, T. E. and Brown, G. H. 1976 and 1984.

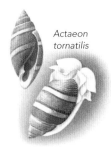

*Actaeon*
*tornatilis*

## ORDER BULLOMORPHA
### BUBBLE SHELLS
Shell fragile, external, internal or sometimes absent; mantle may be extensive, enfolding the shell, or reduced; gills internal; foot sole-like for creeping or extended into flaps for swimming.

### Actaeon tornatilis (Linnaeus)
**Shell** about 20mm long and barrel-like with up to 7 whorls; resembles a prosobranch (see pages 146–161).
**Colour** pink-grey-yellow.
**Habitat** burrowing in sand or mud, generally on the lower shore and in shallow water.
**Distribution** Mediterranean, Atlantic, English Channel and North Sea.

*Bulla*
*striata*

### Bulla striata (Brugière) ( =Bullaria striata)
**Body** up to 60mm long and unable to withdraw into the shell.
**Shell** about half as long as body; spire reduced.
**Colour** shell brownish, patterned.
**Habitat** on sand and mud, often among seaweeds on

the lower shore and in shallow water.
**Distribution** Mediterranean, and Atlantic.

*Scaphander
lignarius*

### *Scaphander lignarius* (Linnaeus)
**Body** may reach 140mm long and is unable to withdraw into the shell.
**Shell** about half as long as body; aperture tapers towards the apex.
**Colour** shell yellow-white.
**Habitat** on sandy and muddy substrates.
**Distribution** Mediterranean and Atlantic north to English Channel.

whole
animal

### *Haminea hydatis* (Linnaeus)
Not resembling a prosobranch; animal can withdraw completely into shell.
**Shell** up to 15mm high, not quite so wide, fragile; aperture height greater than spire height; aperture large and eel-like.
**Animal** up to 30mm long; discordal head carries flat tentaculate processes which hide the front of the shell, the rear being covered by the hind flap of the mantle; comb-like sense organs with up to 12 pairs of lamellae lie on either side of the head.
**Colour** shell translucent yellow-white; animal brownish.
**Habitat** creeping on muddy sand and weeds from the shore down, sometimes swimming.
**Distribution** Mediterranean and Atlantic north to Ireland and Wales, west English Channel. NB: several similar species, including *H. navicula* (da Costa).
**Shell** up to 32mm and body up to 70mm; comb-like sense organs with about 20 pairs of lamellae; never swimming.

*Haminea
hydatis*

### *Philine aperta* (Linnaeus)
**Body** up to 20mm long; enclosing the shell.
**Shell** thin, reduced.
**Colour** body grey-white and translucent; shell white.
**Habitat** on sand on the extreme lower shore and in shallow water.
**Distribution** Mediterranean, Atlantic, English Channel, North Sea and west Baltic.

*Philine
aperta*

# ORDER PLEUROBRANCHOMORPHA

The shell is internal or lacking, and the body is covered dorsally by a membranous shield.

*Pleurobranchus membranaceus*

### *Pleurobranchus membranaceus* Montagu ( =*Oscanius tuberculatus*)

**Body** up to 120mm long; dorsal shield from under which protrudes broad red foot, 2 head tentacles and a gill on the right side.
**Colour** orange-yellow-white.
**Habitat** on mud and gravel in shallow water.
**Distribution** Mediterranean, Atlantic and English Channel.

### *Berthella plumula* (Montagu)

**Shell** hidden by mantle, reaching up to 30mm and animal up to 60mm; shell is much larger in relation to body.
**Colour** body pale yellow or orange, sometimes with patterns on the shield-like mantle.
**Habitat** on the shore in pools and in shallow water down to 10m, feeding on compound sea-squirts etc.
**Distribution** Mediterranean and Atlantic north to south-west Britain.

*Berthella plumula*

Shell only

# ORDER APLYSIOMORPHA

Shell internal or absent; animal slug-like; head long with anterior oral tentacles and posterior tentacles (rhinophores); body with reduced mantle, mantle cavity open; parapodial flaps and large foot. The precise identification may be difficult: see Thompson T.E. 1976 or Thompson T.E. & Brown G.H. 1976.

### *Aplysia punctata* Cuvier
### Sea Hare

**Shell** internal, covered by the mantle and up to 40mm long; very delicate; aperture wide.
**Animal** body up to 200mm long and generally slug-like; at the sides extended up into two flaps of parapodia which join together high up at the rear; head with two pairs of tentacles, the second pair (rhinophores) somewhat enrolled and tube-like, the first pair with leaf-like edges.
**Colour** shell yellowish-amber; animal purple-brownish or olive-green.
**Habitat** in shallow water, especially in spring and summer.
**Distribution** Mediterranean and Atlantic north to Norway. N.B. when disturbed can release purple dye into the water. Similar species include *A. depilans* and *A. fasciata* (opposite).

*Aplysia punctata*

### *Aplysia depilans* Gmelin
**Sea Hare**
Very similar to *A. punctata* (opposite); the body flaps or parapodia are joined together high up; sole of creeping foot is quite wide so that the foot may appear sucker-like; animal up to 300mm long.
**Colour** brown to green with darker marks.
**Habitat** in shallow water, sometimes swimming.
**Distribution** Mediterranean and Atlantic north to south-west Britain. N.B. colour removed to highlight structure.

*Aplysia depilans*

### *Aplysia fasciata* Poiret
**Sea Hare**
Like the two preceding species; however the body flaps or parapodia are not joined together high up at the rear, but low down, and appear well separated both at the front and rear. Body up to 40mm.
**Colour** dark brown-black; parapodia sometimes have red borders.
**Habitat** as above, often swimming.
**Distribution** Mediterranean and Atlantic north to south-west Britain. N.B. colour removed to highlight structure.

*Aplysia fasciata*

## ORDER SACOGLOSSA

A shell is lacking in most species. These animals are often highly coloured.

### *Elysia viridis* (Montagu)
**Body** about 30mm long; flattened and soft: 2 head tentacles; no gills.
**Colour** green.
**Habitat** on green seaweeds such as *Codium* (see page 25) from the middle shore downward.
**Distribution** Mediterranean, Atlantic, English Channel, North Sea and west Baltic.

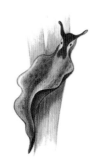

*Elysia viridis*

# ORDER NUDIBRANCHIA

Slug-like opisthobranchs lacking a shell; often highly coloured; external gills of feathery form arranged on the back near to or surrounding the anus; flanks sometimes carry groups of tentaculate defence organs called cerata; head may be clearly visible, carrying smooth oral tentacles and prominent rhinophores divided towards their tips into feathery plates; the front of the foot may have parapodial tentacles. For more information reference should be made to Hunnam, P. and Brown, G. 1975, Thompson, T. E. 1976, Thompson, T. E. and Brown, G. H. 1976 and Thompson, T. E. and Brown, G. H. 1978.

*Tritonia hombergi*

### Tritonia hombergi Cuvier

**Body** up to 200mm in length (the largest nudibranch found in north-west European waters), distinctly bilobed oal veil at the front of the head bearing a margin of papillae. Thick tubular sheath-like bases to the head tentacles (known as rhinophores) with many feathery branches protruding. Branching gills along the side of the body.
**Colour** young forms are whitish and translucent becoming brownish-purple with age on the upper surface.
**Habitat** shallow and deeper water down to 80m, often associated with *Alcyonium digitatum* (see page 101), on which it feeds.

*Dendronotus frondosus*

### Dendronotus frondosus (Ascanius)

**Body** about 100mm long; head tentacles are branched and there are many branched appendages arranged in pairs along the back.
**Colour** yellow-pink-white with darker markings on the back.
**Habitat** among rocks and on sand in shallow water and down to about 100m.
**Distribution** Atlantic, English Channel, North Sea and west Baltic.

### Doto pinnatifida Trinchese ( =Doto splendida)

**Body** up to 30mm long; quite narrow; head low and flat with slight frontal flaps; conspicuous pair of rhinophores surrounded by sheaths that carry dark brownish dots.
**Colour** body light to dark brown, carrying up to 9 pairs of cerata; these carry branches arranged in rings; the the branch tips are spotted dark brown or black; black-tipped warts along the sides of the body.
**Habitat** in shallow water, often among hydroids e.g. *Nemertesia antennina* (see page 83).
**Distribution** Atlantic, western English Channel.
**Similar species** five including *Doto coronata* Gmelin (also illustrated), which is up to 15mm long, often less; with conspicuous side flaps on the front of the head; rhinophore sheaths open; white to yellow or pink with reddish or purple spots or marks at the base of each ceras; 8 pairs of cerata which bear branches arranged in rings; branch tips red.

*Doto coronata*

*Doto pinnatifida*

### Acanthodoris pilosa (Müller)

**Body** may reach up to 60mm long but often less; nearly half as wide; front end rounded; oral tentacles usually concealed when viewed from above; a pair of rhinophores with feathery tips; mantle surface covered with soft tubercles; a circlet of 9 large branching gills surrounds the anus.
**Colour** whitish-grey-brownish, even to purple.
**Habitat** on the lower shore and in shallow water down to 80m.
**Distribution** Mediterranean, Atlantic, English Channel and North Sea.

*Acanthodoris pilosa*

*Polycera quadrilineata*

### Polycera quadrilineata (Müller)

**Body** up to 30mm long, quite narrow; head having 2 pairs of smooth narrow tentacles with orange tips and 1 pair of thicker rhinophores with orange-yellowish tips; up to 11 branching gills with yellow tips surround the anus and these lie between a further pair of smooth orange-tipped tentacles.
**Colour** body white with patches of yellow and sometimes black marks.
**Habitat** shallow water down to 200m.
**Distribution** Mediterranean, Atlantic, English Channel and North Sea.

### Greilada elegans Bergh

**Body** up to 40mm long, quite narrow; conspicuous rhinophores; front edge of the mantle carries numerous small projecting processes, these lacking towards the rear; 5–7 branching gills just anterior to the anus.
**Colour** unmistakeable: orange-yellow with conspicuous blue markings on the mantle and sides.
**Habitat** in shallow water down to 25m, often associated with bryozoans such as *Bugula* (see pages 247–248) on which it feeds.
**Distribution** Mediterranean north to south-west Britain and Ireland.

### Palio dubia (Sars)

**Body** up to 30mm long, often less, quite narrow; a pair of slender tentacles occurs below the head; relatively short rhinophores; front edge of mantle carries many rounded yellow tubercles and the sides of the body are warty; more tubercles occur behind on either side of the gill circlet; 3–5 branching gills present.
**Colour** yellow to green.
**Habitat** in rock pools and down to 100m, usually feeding on ectoprocts.
**Distribution** Mediterranean, Atlantic, English Channel and North Sea.

*Palio dubia*

*Greilada elegans*

167

*Thecacera pennigera*

### Thecacera pennigera (Montagu)

**Body** up to 30mm long, quite narrow; anterior end of foot bearing a pair of small flat parapodial tentacles; head carries rhinophores borne in sheaths, the lips of which are drawn out into tubercle behind; 3–5 branching gills lie around the anus and posterior to these lie a pair of conspicuous club-like tentacles.
**Colour** white with many small black spots interspersed with larger orange ones.
**Habitat** in shallow water down to 20m, often associated with the bryozoan *Bugula* on which it feeds.
**Distribution** Mediterranean, Atlantic north to south-west Britain and English Channel.

### Limacia clavigera (O. F. Müller)

**Body** about 20mm long; flat with over 20 appendages of varying sizes; these appendages are difficult to distinguish from the head tentacles; 13 branched gills around the anus.
**Colour** body white, appendages usually have orange-red tips.
**Habitat** usually in shallow water.

*Limacia clavigera*

### Cadlina laevis (Linnaeus)

**Body** about 20mm long; small warts; 2 unbranched head tentacles and 5 branching gills in a ring on the back.
**Colour** pure white with pale yellow spots on the sides.
**Habitat** in pools on the lower shore and in shallow and deeper water.
**Distribution** Atlantic north from Scotland and North Sea.
**Distribution** Mediterranean, Atlantic, English Channel and North Sea.

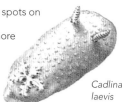

*Cadlina laevis*

### Rostranga rubra (Risso) ( =R. rufescens)

**Body** up to 15mm long, a little less than half as wide; front end blunt; head and finger-like oral tentacles not visible from above; yellowish rhinophores each with about 12 pairs of plate-like branches; mantle bearing short stubby tubercles; 10 branching gills surround the anus.
**Colour** scarlet, occasionally yellowish, with a scattering of black spots.
**Habitat** among sponges the colour of which it often matches, in shallow water.
**Distribution** Mediterranean and Atlantic coasts north to Ireland and Scotland.

*Rostranga rubra*

### Archidoris pseudoargus (Rapp)
#### Sea-lemon

**Body** up to 70mm long; 2 unbranched head tentacles; back covered with small warts and bearing anus towards rear; 9 branching gills in a ring on the back.
**Colour** yellowish with brown-green-pink markings.

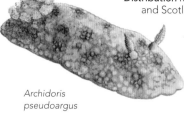

*Archidoris pseudoargus*

Habitat moves up to the lower shore in Summer to spawn, otherwise in deeper water, generally among rocks.
Distribution Atlantic, English Channel and North Sea.

### Jorunna tomentosa (Cuvier)

Similar to *A. pseudoargus* (above).
Body up to 40mm long; small warts on the back, 2 brownish unbranched head tentacles and 15 whitish branching gills in a ring on back.
Colour yellowish with brown markings.
Habitat among rocks on the lower shore in Summer and in deeper water at other times.
Distribution Mediterranean, Atlantic, English Channel and North Sea.

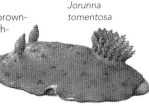

*Jorunna tomentosa*

### Aeolidia papillosa (Linnaeus)
#### Common Grey Sea-slug

Body may be 80mm long; 2 pairs of unbranched head tentacles; many appendages carried on the back which are 'parted' in the middle.
Colour grey-brown.
Habitat on stony and rocky shores, between high and low water marks.
Distribution Atlantic, English Channel and North Sea.

*Aeolidia papillosa*

*Spurilla neapolitana*

### Spurilla neapolitana (Della Chiaje)

Body up to 60mm long, less than half as broad, tapering to tail; conspicuous head carries oral tentacles and rhinophores; branching gills arranged in pairs along the body.
Colour overall brownish.
Habitat in shallow water.
Distribution Mediterranean and Atlantic north to France and possibly south-west Britain.

### Crimora papillata Alder & Hancock

Body up to 35mm long, a little less than half as wide; front end rounded; tapering; most of the anterior and lateral margins carry elaborately divided yellow-orange tubercles which are better developed at the front; orange rhinophores; between 3–5 branching gills surround the anus.
Colour body white to pale yellow.
Habitat in shallow water down to 80m feeding on leafy bryozoa such as *Flustra*.
Distribution until fairly recently this species was thought to be rare in European waters but is now known from the Mediterranean and Atlantic north to south-west Britain where it can be found in shallow water down to 30m usually associated with ectoprocts such as *Flustra foliacea* (see page 247).

*Crimora papillata*

*Antiopella cristata*

### Antiopella cristata (Chiaje)
**Body** up to 75mm long, quite narrow, gradually tapering to tail. The head is not easily made out from above and carries two oral tentacles; rhinophores quite well developed and almost joined at their bases; mantle bearing many conspicuous cerata, the tips of which have small white dots.
**Colour** yellow, pale brown or creamy.
**Habitat** often found near ectoprocts *Bugula* and *Cellaria* (see pages 247–248).
**Distribution** Mediterranean, Atlantic north to south-west Britain and Ireland, English Channel and northern North Sea.

### Facelina auriculata (O. F. Müller)
**Body** up to 25mm long; thin; 2 dissimilar pairs of unbranched head tentacles and about 6 groups of appendages on the animal's back.
**Colour** pale; appendages dark red with white tips.
**Habitat** among rocks on the lower shore and in shallow water.
**Distribution** Atlantic and English Channel.

*Facelina auriculata*

### Facelina bostonensis (Couthouy)
(formerly included in *F. auriculata*)
**Body** up to 50mm long, may be less, fairly narrow; head with long narrow oral tentacles, slightly shoreter rhinophores and flattish propodial tentacles; many cerata arranged on either side of the body in clusters of 8, each contains a brownish or greenish pigment strip and a white band near the tip; no iridescent sheen.

*Facelina bostonensis*

**Colour** body whitish with traces of pink.
**Habitat** among rocks on the shore and in shallow water; often associated with hydroids e.g. *Tubularia* (see page 78).
**Distribution** Mediterranean, Atlantic, western English Channel and northern North Sea.

*Facelina coronata*

### Facelina coronata (Forbes and Goodsir)
**Body** up to 38mm long; very similar to *F. bostonensis* (above) but even narrower.
**Colour** white with pinkish tinge and normally a bluish iridescence in the cerata, which have an underlying pinkish-red strip.
**Habitat** as for *F. bostonensis* (above).
**Distribution** Mediterranean, Atlantic, western English Channel and northern North Sea.

### Coryphella pedata (Montagu)
**Body** up to 40mm long, quite narrow and tapering towards the rear; head clearly visible, with conspicuous oral tentacles and

rhinophores; the mantle bears cerata grouped along the dorsal aspects of the flanks in clusters of about 6.
**Colour** each of the cerata contains a bright orange spot and has a white tip; body violet.
**Habitat** among hydroids down to about 40m. Often associated with the hydroid *Eudendrium ramosum* (see page 80).
**Distribution** Mediterranean, Atlantic north to south-west Scotland, Ireland, English Channel and northern North Sea.

### *Coryphella lineata* (Loven)
**Body** up to 40mm long, much narrower; head clearly visible, bearing conspicuous tapering oral tentacles and tapering rhinophores; either side of the mantle extended into numerous cerata arranged in clusters of up to five.
**Colour** background translucent-whitish with more opaque patches and a conspicuous midline which runs up to the bases of the oral tentacles, where it branches to ascend each; such a line also passes down each side and joins the midline about half-way along the body; there is a similar one on the back of the rhinophores; each of the cerata contains a bright red mark and an opaque white tip.
**Habitat** among hydroids down to quite deep water. Often associated with the hydroid *Tubularia indivisa* (see page 78).
**Distribution** Mediterranean and Atlantic north to Scotland, English Channel and North Sea.

*Coryphella pedata*

*Coryphella lineata*

### *Caloria elegans* (Alder and Hancock)
**Body** about 35mm long, fairly narrow; head with conspicuous oral tentacles, each bearing a fine white stripe, and a pair of rhinophores; body bearing many cerata with an orange-red core, a dark subterminal band and a white tip.
**Colour** background colour greyish-brown.
**Habitat** in shallow water.
**Distribution** Mediterranean, Atlantic north to France and south-west Britain.

*Caloria elegans*

### *Cuthona caerulea* (Montagu)
**Body** up to 18mm long; head clearly visible and bearing long narrow smooth tentacles and rhinophores; the cerata are arranged in up to 10 rows across the body.
**Colour** body white, sometimes slightly greenish; cerata characteristically pigmented with a subterminal blue band and a terminal orange tip; sometimes there is an inner orange band on the body side of the blue one.
**Habitat** in shallow water among hydroids.
**Distribution** Mediterranean, Atlantic north to Norway and North Sea.

*Cuthona caerulea*

*Limnaea stagnalis*

# SUBCLASS PULMONATA

Gastropods which undergo torsion and bear a shell. The operculum is lacking. The mantle cavity is modified to form a lung for breathing air, and may be further adapted to permit respiration under water. Although the pulmonates are generally terrestrial or freshwater animals, several species fall within the scope of this book.

### *Limnaea stagnalis* Linnaeus
**Great Pond Snail**
**Shell** up to 50mm high, but not as wide; high spire with about 6 quite well-defined whorls; horn-like.
**Colour** pale-mid-brown.
**Habitat** in fresh water, in ponds and ditches among weeds throughout Europe; in marine areas where the salinity is very low, usually in sheltered places among seaweeds.
**Distribution** Baltic.

### *Limnaea peregra* O. F Müller
**Wandering Pond Snail**
**Shell** up to 20mm high, but not as wide; oval; spire not sharply pointed and with about 5 quite well-defined whorls.
**Colour** pale-mid-brown.
**Habitat** in fresh water, dwelling in swamps and bogs in many parts of Europe; seldom found in rivers; in marine areas where the salinity is very low, usually in sheltered and exposed areas among stones and seaweeds.
**Distribution** Baltic. N.B. this species is very variable in appearance.

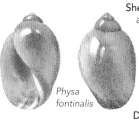

*Limnaea peregra*

### *Physa fontinalis* Linnaeus
**Bladder Snail**
**Shell** up to 12mm high, but not as wide; short spire with about 4 quite well-defined whorls; when crawling, 2 mantle lobes show outside the shell, the anterior lobe with 9 finger-like projections and the posterior one with 6; horn-like, transparent and delicate.
**Habitat** in fresh water; often associated with water weeds such as *Potamogeton sp.*; more abundant in northern Europe; in marine areas where the salinity is very low.
**Distribution** Baltic.

*Physa fontinalis*

*Phytia myosotis*

### *Phytia myosotis* Draparnaud ( =*Alexia myosotis*)
**Shell** about 9mm high and about half as wide; about 7 whorls; aperture may show 3 ridges on the inner lip and 1 on the outer.
**Colour** grey-yellow-brown.
**Habitat** in estuaries, salt marshes and under stones on the upper shore.
**Distribution** Mediterranean, Atlantic and English Channel.

### *Leucophytia bidentata* (Montagu)
**Shell** about 9mm high and about half as wide; conical spire; about 6 whorls; aperture shows 2 ridges on the inner wall.
**Colour** grey-white.
**Habitat** in crevices and among seaweed debris on the upper shore and in salt marshes.
**Distribution** Atlantic and English Channel.

*Leucophytia bidentata*

### *Otina ovata* (Brown) *( =Otina otis)*
**Shell** up to 5mm high and about half as wide; ear-shaped; about 2 whorls; animal and shell slightly resemble a minute ormer when creeping over rocks.
**Colour** red-brown-purple.
**Habitat** in crevices and old barnacle shells on the upper shore.
**Distribution** Atlantic and English Channel.

*Otina ovata*

# CLASS SCAPHOPODA

This is a small class of bilaterally symmetrical molluscs. The three-lobed foot is somewhat reduced and projects from the wider end. It is used for burrowing. The shell is tubular and tapering. The head is also reduced. These animals live partly buried in sand and mud, with the wide end of the shell (head end) pointing downwards. About 350 species are known from the world's seas and oceans.

### *Dentalium entalis* Linnaeus
**Tusk Shell**
**Shell** up to 50mm long.
**Colour** yellow-white.
**Habitat** sand and mud in deeper water.
**Distribution** Atlantic, English Channel and North Sea.

*Dentalium entalis*

# Class Bivalvia
## ( =Lamellibranchia or Pelecypoda)

These molluscs are bilaterally symmetrical. The body is compressed laterally and enclosed by a shell of two valves, which are linked dorsally by a ligament and hinge. The head is rudimentary or missing, and tentacles and a radula are lacking. The foot is ventral and without a crawling surface. Bivalves are filter and deposit feeders, with the sexes usually separate, and the larva frequently free-living and planktonic. Many are efficient burrowers.

Some important features of bivalve shells are shown in fig. 28, page 143. The two valves may be massive and strong or fine and delicate. The elastic ligament can be inside the shell, outside it, or both. It forces the valves open, but in life its action is opposed by the contractions of the closing or adductor muscles whose attachment scars can often be seen inside the shell. The hinge prevents the two valves slipping out of alignment with each other, yet allows them to open and close. It may be smooth, crenulate or toothed. Each valve develops from the beak above which lies the convex *umbo* (pleural umbones). The margin or free shell edge may also be smooth, crenulate or toothed, and the outer surface of the valves can be of many textures. Sometimes part of the periostracum (see page 143) persists over it. In many burrowing species the mantle (also discussed on page 143) is extended posteriorly via a gap in the valves as two siphons which enable the animal to draw in fresh sea water bearing food and oxygen, even though it may be buried in sand itself. The foot is often used for burrowing, but sometimes secretes threads (byssus) which are used for sticking the animal to rocks. In the illustrations that follow almost all the bivalves are shown with their anterior ends pointing towards the right-hand side, and their dorsal surfaces uppermost. For a more detailed account of many European bivalves, see Tebble, N. 1976. About 15,000 species of bivalve are known worldwide from all aquatic habitats.

*Nucula nucleus*

### *Nucula nucleus* (Linnaeus)
**Common Nut-shell**
**Shell** up to 12.5mm long; valves similar; edge crenulate; hinge with more anterior teeth than posterior teeth; adductor scars equal.
**Colour** periostracum brown-green-yellow; dark brown internal ligament.
**Habitat** in clay, gravel and sand from shallow water down to about 150m.
**Distribution** Mediterranean, Atlantic, English Channel, North Sea and west Baltic. N.B. animal lacks siphons.

### *Arca noae* Linnaeus
**Noah's Ark Shell**
**Shell** up to 80mm long; valves similar; edge smooth, apart from crenulate posterior; straight hinge with many small, equal teeth; external ligament; adductor scars equal; dorsal view shows umbones far apart; outer surface ribbed and sometimes covered with short-haired periostracum.

*Arca noae*

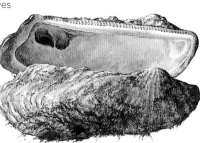

lateral view

**Colour** dark brown with lighter marks; periostracum brownish.

dorsal view

**Habitat** attached to rocks and stones by byssus threads, usually offshore.
**Distribution** Mediterranean and Atlantic.

### *Arca tetragona* Poli
**Cornered Ark Shell**
**Shell** up to 50mm long; similar to *A. noae* (above) but valves more rectangular and finely sculptured; may be encrusted with other organisms.
**Colour** periostracum brown; exterior white-yellow, interior white with darker patches.
**Habitat** on stones and rocks (to which it is attached by a green byssus) on lower shore and down to about 100m.
**Distribution** Mediterranean, Atlantic and English Channel.

dorsal view

*Arca tetragona*

lateral view

### *Glycymeris glycymeris* (Linnaeus)
**Dog Cockle**
**Shell** up to 65mm long and nearly circular; valves similar; edge crenulate; 2 rows of up to 12 teeth on each hinge; external ligament; umbones separated but not so greatly as in the above shells; outer surface finely sculptured.
**Colour** typical brown markings on exterior, interior white or brown.
**Habitat** burrowing just below the surface of mud, sand or gravel from shallow water down to about 80m.
**Distribution** Mediterranean, Atlantic, English Channel and Baltic.

*Glycymeris glycymeris*

175

right valve

*Anomia ephippium*

left valve

### *Anomia ephippium* Linnaeus
**Common Saddle Oyster**
**Shell** up to 60mm long; thin, flat, lower (right) shell with aperture; thicker, domed, upper (left) valve; 1 adductor and 2 byssus muscle scars on upper valve, 1 adductor scar on lower; upper valve with scaly outer surface often covered with other organisms.
**Colour** white-pale brown.
**Habitat** attached to rocks and other shells (to whose shapes it often conforms), from middle shore downward.
**Distribution** Mediterranean, Atlantic, English Channel and North Sea. N.B. calcified byssus connects the upper valve to the substrate via the aperture in the lower valve. Animal almost circular in outline.

### *Mytilus edulis* Linnaeus
**Common Mussel**
**Shell** varies in length from 10–100mm; typical mussel shape with similar valves; edge smooth; hinge lacks conspicuous teeth but has up to 12 crenulations near the umbones; beaks terminal; ligament external; anterior adductor scar small, posterior scar large.
**Colour** exterior brown-blue-black, sometimes with brown markings; perostracum thin and dark brownish; periphery of mantle white-yellowish; interior pearly with dark border.
**Habitat** on stones and rocks in estuaries and on rocks on more exposed shores often in extensive beds and associated with barnacles, from middle shore downward.
**Distribution** Mediterranean, Atlantic, English Channel, North Sea and Baltic.

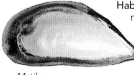

*Mytilus edulis*

### *Mytilus galloprovincialis* (Lamarck)
**Mediterranean Mussel**
For a long time this has been regarded as a separate species, but there is debate as to whether it is a race of *M. edulis* (above) which it closely resembles apart from the following differences.
**Shell** umbones more pointed and turned down; broader and less angular dorsally.
**Colour** periphery of mantle dark.
**Habitat** not found in estuaries.
**Distribution** Mediterranean and Atlantic north to south-west England, south Wales and southern Ireland.

*Mytilus galloprovincialis*

### *Modiolus modiolus* (Linnaeus)
**Horse Mussel**

*Modiolus modiolus*

**Shell** up to 140mm long, but may reach 200mm; valves similar; edge smooth; hinge without teeth; beaks not terminal, but a little way from the anterior; anterior adductor scar small, posterior scar large; mantle margin not frilled; shell thick with horny periostracum which may have spines in young individuals.
**Colour** animal usually dark orange; exterior of shell purplish; interior pale.
**Habitat** extreme lower shore down to about 150m, often associated with laminarians (see pages 32–33).
**Distribution** Atlantic from Biscay northward, English Channel and North Sea.

### *Modiolus barbatus* (Linnaeus)
**Bearded Horse Mussel**

*Modiolus barbatus*

**Shell** up to 60mm long, but usually smaller; similar to *M. modiolus* (above) but thick, horny periostracum persists over posterior part and it is arranged in the form of many semicircular rows of serrated whiskers.
**Colour** similar to *M. modiolus*.
**Habitat** on rocks and shells from the lower shore down to about 100m.
**Distribution** Mediterranean, Atlantic, English Channel and North Sea.

### *Musculus discors* (Linnaeus)

**Shell** up to 12.5mm long; valves similar; edge smooth except where the ribs meet it; up to about 12 anterior ribs, about three times as many posterior, often smaller; beaks not quite terminal.
**Colour** shell paler than preceding species.
**Habitat** under rocks and among seaweeds and other invertebrates from the middle shore downward.
**Distribution** Mediterranean, Atlantic, English Channel and North Sea.

*Musculus discors*

### *Brachyodontes minimus* (Poli)
**Dwarf Mussel**

**Shell** up to 10mm long; valves similar and very thin; edge smooth; hinge with minute teeth; terminal umbones sometimes worn away; horny periostracum.
**Colour** exterior dark brown with violet marks; interior pearly.
**Habitat** usually in shallow water attached by byssus threads to rocks, etc.
**Distribution** Mediterranean and Atlantic.

*Brachyodontes minimus*

177

### *Pteria hirundo* (Linnaeus)
**Wing Oyster**

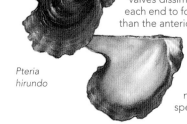

*Pteria hirundo*

Shell up to 75mm long, unusual asymmetrical shape; valves dissimilar; edge smooth; dorsal region drawn out at each end to form 2 ears, posterior ear up to six times longer than the anterior ear.

Colour periostracum brown; exterior grey-brown; interior pearly white.

Habitat attached to stones in mud, clay and gravel, sometimes in quite deep water.

Distribution Mediterranean and Atlantic north to the British Isles. N.B. several related species occur in these regions.

### *Pinna fragilis* Pennant
**Fan Mussel**

Shell up to 300mm long; fan-shaped; similar valves; edge smooth, though sometimes broken; hinge lacks teeth; external ligament; anterior adductor scar small, posterior scar larger; outer surface has concentric lines and ribs, occasionally with spines.

Colour exterior brownish; interior brownish and glassy.

Habitat standing upright in muddy sand and gravel attached to a sunken stone or pebble by byssus threads, often in quite deep water.

Distribution Atlantic and English Channel.

### *Pinna nobilis* Linnaeus
**Fan Mussel**

Similar to *P. fragilis* (above).
Shell up to 450mm long; outer surface of shell has prominent overlapping scales.
Colour exterior red-brown; interior violet-blue-grey.
Habitat as for *P. fragilis*.
Distribution Mediterranean and Atlantic.

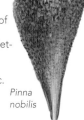

*Pinna nobilis*

*Pinna fragilis*

### *Ostrea edulis* Linnaeus
**Common European Oyster** or **Flat Oyster**

Shell up to 100mm long; shape rounded and very variable; valves dissimilar; lower valve (left) saucer-like, prominently sculptured and often attached to rocks and stones; upper (right) flat and sculptured; margin often crenulate; hinge lacks teeth; internal ligament, single adductor scar; periostracum very thin.

Colour grey-brown.

Habitat from shallow water down to about 80m where there is a suitable substrate, and in commercial beds.

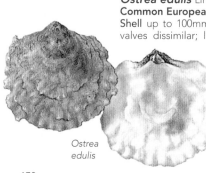

*Ostrea edulis*

Distribution Mediterranean, Atlantic, English Channel and North Sea.

## *Crassostrea angulata* (Lamarck)
**Portuguese Oyster**
Shell up to 150mm wide and about half as long; sculptured valves dissimilar; lower valve (left) trough-like; upper (right) flattish; edges usually interfold; hinge lacks teeth; internal ligament.
Colour periostracum dirty brown; exterior dirty white; interior white-purple.
Habitat in shallow water on rocks and stones and in commercial beds.
Distribution Atlantic, English Channel and North Sea.

## *Pecten maximus* (Linnaeus)
**Great Scallop** or **St James' Shell**
Shell up to 150mm long; upper (left) valve flat, conspicuous ribs are rounded in section.

*Crassostrea angulata*

Colour upper valve red-brown; lower (right) is white-brown with brownish markings.
Habitat on sand and gravel, usually in quite deep water.
Distribution Atlantic, English Channel and North Sea. N.B. the illustration shows the interior of the upper valve and the exterior of the lower one.

## *Pecten jacobaeus* Linnaeus
**Fan-shell**
(Not illustrated)
Shell up to 130mm long; typical scallop-shape with conspicuous ribs and ears; upper (left) valve is flat; lower (right) is saucer-like and ribs on shell are not rounded but more square in section.
Colour upper valve red-brown and occasionally spotted, lower valve pinkish.
Habitat on sand and gravel, usually in quite deep water.
Distribution Mediterranean.

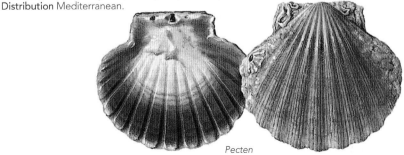

Left valve     *Pecten maximus*     Right valve

179

*Chlamys varia*

### *Chlamys varia* (Linnaeus)
**Variegated Scallop**
**Shell** up to 60mm long, slightly wider; oval; both valves convex but not exactly similar; about 28 ribs bearing scale-like teeth which are usually abraded near the umbones and may be rubbed from most of the shell; edge indented by ribs; hinge lacks teeth in adult, hinge length is about half shell width; one adductor scar towards posterior; posterior 'ear' half to one-third length of anterior.
**Colour** very variable; purple-red-white-yellow-brown, sometimes patterned.
**Habitat** living free or attached by byssus threads to the substrate, extreme lower shore down to 80m.
**Distribution** Mediterranean, Atlantic, English Channel and North Sea.

### *Aequipecten opercularis* (Linnaeus)
**Queen Scallop**
**Shell** up to 90mm long; rounded; similar to *C. varia* (above) but lower (right) valve less convex than upper (left); about 20 ribs; anterior ear slightly longer than posterior one.
**Colour** very variable as in *C. varia*; sometimes spotted or striped.
**Habitat** attached to substrate by byssus threads when young, usually on gravel and sand, occasionally on extreme lower shore and down to about 200m.
**Distribution** Mediterranean, Atlantic, English Channel and North Sea. N.B. this scallop can swim very actively by flapping its shells.

*Aequipecten opercularis*

### *Palliolum tigerina* (O. F. Müller)
**Tiger Scallop**
**Shell** up to 25mm long; rounded; valves almost similar, lower (right) fractionally less convex than upper (left); sometimes smooth, sometimes with many fine ribs and sometimes with a few prominent ones; posterior ear very much smaller than anterior one.
**Colour** brown-yellow-white, being variable and sometimes patterned.
**Habitat** on sand, gravel and stones from extreme lower shore down to about 100m.
**Distribution** Atlantic, English Channel and North Sea.

*Palliolum tigerina*

### *Chlamys distorta* (da Costa)
**Hunchback Scallop**
**Shell** up to 40mm long, slightly deeper; outline oval; ears not quite equal; outside of shell sculptured by about 70 fine ribs.
**Habitat** attached by byssus when young, then living cemented to hard substances by the right valve from the shore down to 100m.
**Distribution** Mediterranean, Atlantic north to Norway, English Channel and North Sea.

*Chlamys distorta*

### *Lima hians* (Gmelin)
**Gaping File-shell**
**Shell** up to 25mm long; asymmetrical; with about 50 spiny ribs; when viewed from the front there is a very conspicuous gape between the valves.
**Colour** delicate white, becoming dirtier and browner in older specimens.
**Habitat** from extreme lower-shore down to about 100m, sometimes in a 'nest' made from stones constructed by using its byssus threads, and among holdfasts of *Laminaria* (see pages 32–33).
**Distribution** Mediterranean, Atlantic and English Channel. N.B. very conspicuous, non-retractile, orange-coloured tentacles around the shell edge. This species can swim.

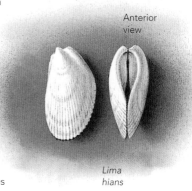

Anterior view

*Lima hians*

### *Astarte sulcata* (da Costa)
**Shell** up to 45mm long but not as wide; solid and heavy; valves similar; edge smooth; hinge with apparently 2 cardinal teeth on each valve; beaks slightly anterior of centre; umbones facing forward; external ligament; anterior and posterior adductor scars similar.
**Colour** black-brown periostracum covers white, finely and concentrically lined exterior; interior pale brown-white.
**Habitat** burrowing in muddy seabeds.
**Distribution** Mediterranean, Atlantic and North Sea, north to southern Arctic. N.B. animal lacks siphons and has a very long foot which may be extended if left undisturbed.

*Astarte sulcata*

### *Loripes lucinalis* (Lamarck)
**Shell** up to about 20mm long, rather thin, rounded; valves similar; edge smooth; hinge with 2 small cardinal teeth and 2 minute lateral teeth pervalve; umbones roughly central and pointing anteriorly; external ligament; posterior adductor shorter and rounder than anterior; periostracum reduced; outer surface has five concentric lines.
**Colour** exterior yellow-white; interior white.
**Habitat** burrowing in clay, sand and gravel from extreme lower shore down to about 150m.
**Distribution** Mediterranean, Atlantic, English Channel and North Sea.

*Loripes lucinalis*

*Myrtea
spinifera*

### Myrtea spinifera (Montagu)

**Shell** up to 25mm long but not as wide; oval; valves similar; edge smooth; left valve with 2 cardinal and 2 lateral teeth (1 on either side), right valve with 1 cardinal tooth and similar lateral ones; beaks slightly anterior of centre, posterior adductor scar smaller than anterior; periostracum reduced; fine lines on outer surface concentrically arranged.

**Colour** exterior white-cream; interior white.

**Habitat** in silt or muddy gravel on the seabed from about 10m to 100m.

**Distribution** Mediterranean, Atlantic and English Channel.

*Montacuta
ferruginosa*

### Montacuta ferruginosa (Montagu)

**Shell** up to 8mm long but considerably less wide; oval; valves similar; edge smooth.

**Colour** thin, reddish periostracum; interior white-purple.

**Habitat** a common commensal of *Echinocardium cordatum* (see page 265) which burrows in sand on the lower shore and in shallow water.

**Distribution** Mediterranean, Atlantic, English Channel and North Sea.

### Arctica islandica (Linnaeus) ( =Cyprina islandica)

**Shell** up to 125mm long but not as wide; solid, heavy, oval-shaped; valves similar; edge smooth; hinge with 3 cardinal teeth and 1 lateral tooth behind; umbones anterior to middle of shell and pointing forward; external ligament; anterior adductor scar a little smaller than posterior; thick periostracum; outer surface concentrically and finely lined.

**Colour** periostracum yellow-red-brown and shiny; interior white.

**Habitat** burrowing in sand and mud from lower shore downward.

**Distribution** Atlantic, English Channel, North Sea and the west Baltic.

*Arctica
islandica*

### Glossus humanus (Linnaeus) ( =Isocardia cor)
**Heart Shell**

**Shell** up to 100mm long; plump, solid; circular in outline and easy to identify because of the spirally coiled umbones which incline anteriorly and away from the hinge; valves equal; external ligament.

**Colour** dark brown periostracum.

**Habitat** in sand and mud on the seabed from about 10m downward.
**Distribution** Mediterranean and Atlantic.

Side view

## *Acanthocardia tuberculata*
(Linnaeus)
**Rough Cockle**
**Shell** up to 60mm long; valves equal, heavy; sculptured with coarse ribs carrying tubercles which are poorly developed towards the posterior; especially in silty places; empty shells may be washed ashore.
**Colour** exterior white-brown; interior white.
**Habitat** on seabed attached to rocks.
**Distribution** Mediterranean and Atlantic north to English Channel and south-west Britain and Ireland.

*Glossus humanus*

Anterior view

## *Laevicardium crassum* Gmelin
**Norway Cockle**
**Shell** up to 60mm long and about the same depth; valves equal; somewhat asymmetrical in outline; outside of shell ornamented by fine ribs; edges grooved.
**Colour** outside white to yellow.
**Habitat** burrowing in soft substrates in quite shallow and deeper water.
**Distribution** Mediterranean, Atlantic north to Norway, English Channel and North Sea.

*Acanthocardia tuberculata*

## *Acanthocardia aculeata*
(Linnaeus)
**Spiny Cockle** or **Red Nose**
**Shell** up to 100mm long; plump; valves similar; edge strongly toothed corresponding to conspicuous furrows on inside and outside of shell; hinge with 2 cardinal teeth in each valve; 1 posterior and 1 anterior lateral tooth in left valve, anterior cardinal tooth of this valve bigger than posterior tooth, right valve with 1 posterior and 2 anterior lateral teeth; umbones forward from the middle of the shell; external ligament; 2

*Laevicardium crassum*

*Acanthocardia aculeata*

183

adductor muscle scars; about 22 ribs on each valve, each rib bearing a line of spines; spines more developed posteriorly and ventrally, anteriorly they are blunter and bend towards the posterior; outer shell with fine concentric markings.

**Colour** exterior yellow-white, interior white.

**Habitat** in sand, from about 10m downward.

**Distribution** Mediterranean, Atlantic, English Channel and North Sea.

*Acanthocardia echinata*

### *Acanthocardia echinata* (Linnaeus)
**Prickly Cockle**

**Shell** up to 75mm long: similar to *A. aculeata* (page 183) but in the left valve the cardinal teeth are similar in size; rib spines of outer shell have bases which are broad and usually anastomose with those of their neighbours.

**Colour, Habitat** and **Distribution** as for *A. aculeata*.

*Parvicardium papillosum*

### *Parvicardium papillosum* (Poli)

**Shell** up to 12.5mm long, rounded, valves similar; edge deeply crenulated; hinge, teeth and adductor scars much as for *A. aculeata* (page 183); beaks just anterior of centre; external ligament, about 25 tuberculated ribs on each valve.

**Colour** periostracum pale brown; exterior white-grey-yellow, often with red-brown patterns; interior white-pink, smooth not grooved.

**Habitat** soft substrates.

**Distribution** Mediterranean and Atlantic.

*Parvicardium exiguum*

### *Parvicardium exiguum* (Gmelin)
**Little Cockle**

**Shell** up to 12.5mm long; less rounded than *P. papillosum* (above), but edge and hinge are alike; beaks well forward; periostracum thicker; about 21 ribs which show tubercles only anteriorly and ventrally when adult.

**Colour** periostracum brown, exterior brown; interior of shell white-green and smooth.

**Habitat** extreme lower shore down to about 60m on sand and mud, sometimes in estuaries and brackish water.

**Distribution** Mediterranean, Atlantic, English Channel and North Sea.

### *Cerastoderma edule* (Linnaeus)
### ( =*Cardium edule*)
**Common Cockle**

**Shell** up to 50mm long; oval; valves similar; edge crenulate all round, corresponding to grooves which run for a short way inside the shell; external ligament which is as long as about one-third of the shell height; umbones slightly anterior to middle; right valve has 2 cardinal teeth, and 2 anterior and 2 posterior lateral teeth; periostracum reduced.

*Cerastoderma edule*

**Colour** exterior brown; interior white with brown marks.

Habitat on lower shore downward burrowing in mud, sand or gravel, in estuaries and in commercial beds; tolerates salinities between $34^o/_{oo}$ and $20^o/_{oo}$.
Distribution Mediterranean, Atlantic, English Channel and North Sea.

*Cerastoderma edule*

posterior view

## *Cerastoderma glaucum* (Poiret)
## ( =*Cardium glaucum* =*Cardium lamarcki*)
### Lagoon Cockle

Shell usually 30–50mm long; more triangular in outline than *C. edule* (opposite); valves similar; edge crenulate anteriorly, smooth posteriorly and with interior grooves which run nearly to the umbones (i.e. much further in than for *C. edule*); length of external ligament about a quarter of shell height (i.e. relatively shorter than in *C. edule*); umbones slightly anterior of middle; periostracum well developed.

*Cerastoderma glaucum*

Colour exterior pale brown-grey; interior dark-light brown.
Habitat usually in brackish water and lagoons and permanently submerged; burrows in soft sand and mud; will tolerate salinities down to $4^o/_{oo}$.
Distribution Mediterranean, Atlantic, English Channel, North Sea and Baltic to the Gulf of Finland.

## *Cerastoderma hauniense* Petersen & Russell

Shell 8–20mm long; similar to *C. glaucum* (above) but with 1 anterior lateral tooth on right valve (*C. glaucum* has 2).
Habitat normally not burrowing, usually attached to seaweeds by byssus threads.
Distribution Baltic north to Stockholm skärgård.

*Cerastoderma hauniense*

## *Dosinia lupinus* (Linnaeus)
### Smooth Artemis

Shell up to 37.5mm long; rounded; valves similar; edge smooth; hinge with 3 cardinal teeth on both valves, the left valve bearing an extra small tooth anteriorly; umbones anterior to middle; beaks pointing forward and bordering on a heart-shaped depression called the *lunule* (seen in the illustration of the anterior end); external ligament; adductor scars roughly equal; fine concentric lines on outer surface of valves.

right valve

Siphons lack horny coating; can reach up to three times shell length and are united apart from the tip.

*Dosinia lupinus*

Colour periostracum yellowish; interior white.
Habitat burrowing in sand and shell gravel from extreme lower shore down to about 125m.
Distribution Mediterranean, Atlantic, English Channel and North Sea.

anterior view

left valve

185

### *Venus verrucosa* Linnaeus
**Warty Venus**

*Venus verrucosa*

**Shell** up to 63mm long; heavy, rounded; valves similar; inside edge crenulate except posteriorly, and sometimes toothed outside; hinge with 3 cardinal teeth on both valves; umbones forward from centre; beaks bordering lunule (see above); external ligament; adductor scars similar; brown periostracum may be present; outer surface of shell strongly marked with concentric ridges which break up into tubercles towards the posterior.

**Siphons** lack horny coating and are united apart from the tip.

**Colour** exterior yellow-white-grey with brown markings; interior white.

**Habitat** just burrowing in sand or gravel from extreme lower shore down to about 100m.

**Distribution** Mediterranean, Atlantic and English Channel.

### *Timoclea ovata* (Pennant) *( =Venus ovata)*
**Oval Venus**

*Timoclea ovata*

**Shell** up to 20mm long; triangular, rounded; valves similar; edge crenulate apart from opposite the extreme ligament which is rather hidden; hinge with 3 cardinal teeth on both valves, no lateral teeth; umbones anterior and beaks pointing forward to border the lunule (see *Dosinia lupinus,* page 202); adductor scars about equal; up to 50 ribs run up the shell and are intersected by concentric lines.
**Siphons** united.
**Colour** exterior white-pale brown; interior white-orange-purple.
**Habitat** just burrowing in sand or gravel between 3–180m.
**Distribution** Mediterranean, English Channel and North Sea.

### *Clausinella fasciata* (da Costa) *( =Venus fasciata)*
**Banded Venus**

*Clausinella fasciata*

**Shell** about 25mm long; valves similar; edge smooth apart from posterior part; hinge with 3 cardinal teeth on both valves; umbones, beaks, *lumule* and adductor scars all similar to *V. verrucosa* (above); external ligament; exterior with a few pronounced, heavy, concentric ridges, between which are many finer lines.
**Siphons** united.
**Colour** exterior white, yellow or pink with darker rays; interior white-purple.
**Habitat** just burrowing in sand or gravel from 3–100m.
**Distribution** Mediterranean, Atlantic, English Channel and North Sea.

### *Chamelea gallina* (Linnaeus) *( =Venus striatula)*
**Striped Venus**
Shell up to 45mm long; triangular, rounded; similar to *C. fasciata* (opposite) but without the broad, heavy, concentric ridges, and with many finer ridges which are closer together ventrally; posterior of shell more elongated.
**Siphons** united.
**Colour** exterior dirty white, cream or yellow with brown marks.

*Chamelea gallina*

**Habitat** burrowing in sand from lower shore down to about 55m.
**Distribution** Mediterranean, Atlantic, English Channel and North Sea.

### *Venerupis rhomboides* (Pennant)
**Banded Carpet Shell**
Shell up to 60mm long; more oval than in the 'Venus' molluscs; valves similar; edge smooth; umbones and beaks well forward; no lunule; external ligament; outer shell with many concentric, but no radiating, lines.
**Colour** pigment arranged in 4 radiating areas; exterior pink-brown with red-brown marks; interior shiny white.

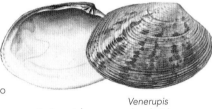

*Venerupis rhomboides*

**Habitat** burrowing in sand and gravel from extreme lower shore down to about 180m.
**Distribution** Mediterranean, Atlantic, English Channel and North Sea.

### *Venerupis pullastra* (Montagu)
**Pullet Carpet Shell**
Shell up to 50mm long; similar shape to *V. rhomboides* (above); outer shell with concentric lines crossed by fine, radiating lines which are more developed posteriorly.
**Colour** exterior cream-grey with patterns; interior shiny white, sometimes with purple marks.
**Habitat** and **Distribution** similar to *V. rhomboides*.

*Venerupis pullastra*

### *Notirus irus* (Linnaeus) *( =Irus irus)*
Shell up to 25mm long; valves similar and longer than their width; edge smooth; lumbones well forward; no lumule; inset external ligament; thin periostracum; outer shell surface bears about 15 concentric ridges with finer radiating markings showing between.
**Colour** exterior dirty white; interior white-yellow.
**Habitat** in holes and crevices of rock and among the holdfasts of *Laminaria* (see pages 32–33), on extreme lower shore and in shallow water.
**Distribution** Mediterranean, Atlantic north to south-west Ireland and west English Channel.

*Notirus irus*

### *Mactra stultorum* (Linnaeus) *( =M. corallina)*
**Rayed Trough Shell**

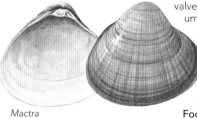

**Shell** up to 50mm long; quite light, triangular; valves similar; edge smooth; hinge complex; umbones almost central; thin external ligament behind umbones, internal ligament as a triangular structure slightly posterior and ventral to the beaks; 2 adductor scars almost equal; fine periostracum; outer shell marked with fine concentric lines.
**Siphons** short and enclosed in a horny sheath.
**Foot** white and pointed.

*Mactra stultorum*

**Colour** brownish rays run from umbones to shell edge; periostracum brown-green; interior purple-white.
**Habitat** burrowing in sand and gravel, from extreme lower shore down to about 100m.
**Distribution** Mediterranean, Atlantic, English Channel and North Sea.

### *Spisula solida* (Linnaeus)
**Thick Trough Shell**
Similar to *M. stultorum* (above).
**Shell** up to 45mm long; more solid; under a hand lens the cardinal teeth can be seen to be finely milled or ridged whereas those of *M. stultorum* are smooth.
**Colour** periostracum light brown; exterior yellow-white with concentric lines and grooves.
**Habitat** similar to *M. stultorum*.
**Distribution** Atlantic, English Channel and North Sea.

### *Spisula elliptica* (Brown)
**Elliptical Trough Shell**
(Not illustrated)
**Shell** up to 46mm long; similar to *S. solida* (above) but lighter and less wide in relation to its length; elliptical; surface almost smooth.

*Spisula solida*

**Habitat** burrowing in sand on lower shore and in shallow water.
**Distribution** Atlantic north from southern Britain, English Channel and North Sea.

### *Lutraria lutraria* (Linnaeus)
**Common Otter Shell**
**Shell** up to 125mm long; oval; valves equal and nearly twice as long as they are wide, and not meeting at either end; edge smooth; hinge complex with 2 cardinal teeth or the left valve forming a conspicuous projection like an inverted v which fits into a similar-shaped socket on the right valve; external and internal ligaments; siphons may extend to more than twice shell length.
**Colour** periostracum brown-green; exterior marked with concentric lines; covered by a transparent sheath which may have brownish markings; interior white.
**Habitat** burrowing in mud, sand and gravel from the extreme

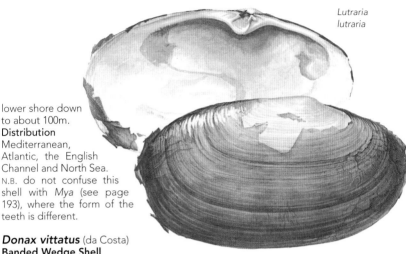

*Lutraria
lutraria*

lower shore down
to about 100m.
**Distribution**
Mediterranean,
Atlantic, the English
Channel and North Sea.
N.B. do not confuse this
shell with *Mya* (see page
193), where the form of the
teeth is different.

### *Donax vittatus* (da Costa)
**Banded Wedge Shell**
**Shell** up to 37.5mm long and half as wide; valves equal; shell
edge strongly toothed; umbones placed well back; external
ligament; shining, polished periostracum; very fine lines radiate
from the umbones.
**Siphons** 2; short.
**Colour** exterior white-yellow-brown-purple, pigment of outer
shell concentrated into bands; interior purple to white.
**Habitat** burrowing in sand from middle shore down to
about 20m.
**Distribution** Mediterranean, Atlantic, English Channel and
North Sea.

### *Callista chione* (Linnaeus) *( =Cytherea chione)*
**Brown Venus**
**Shell** up to 80mm long; outline somewhat triangular; outside
of shell sculptured with
concentric lines.
**Colour** outside of
shell has concentric
yellow to red-
brown lines
alternating with
paler ones, and
radiating bands;
inside pale.
**Habitat** burrow-
ing in sand and
mud down to quite
deep water.
**Distribution** Mediterranean, Atlantic
north to north Wales, western English Channel,
Channel Islands.

*Donax
vittatus*

*Callista
chione*

189

*Angulus
tenuis*

### *Angulus tenuis* da Costa *( =Tellina tenuis)*
**Thin Tellin**
**Shell** up to 20mm long; delicate, flattened, valves almost equal; edge smooth; beaks and umbones slightly posterior; 2 cardinal teeth on each valve, lateral teeth clearly seen on right valve; external ligament; posterior adductor muscle scar shorter than anterior and a little thicker; shiny periostracum; outer surface with very fine concentric lines.
**Siphons** 2; not united, and may be several times as long as the shell.
**Colour** variable; exterior yellow-orange-pink-white with pigment often banded; similar colours inside.
**Habitat** burrowing in sand from the middle shore down to shallow water.
**Distribution** Mediterranean, Atlantic, English Channel, North Sea and Baltic. N.B. owing to the strength of the ligament the 2 valves often remain connected long after the death of the animal, as illustrated. May be present in great numbers.

*Fabulina
fabula*

### *Fabulina fabula* (Gmelin) *( =Tellina fabula)*
**Shell** up to 20mm long; similar to *A. tenuis* but more pointed posteriorly; fine concentric lines and, on the right valve only, very fine diagonal lines and grooves run from left ventral to right dorsal (a hand lens will help here).
**Colour** exterior white-yellow-orange; interior often white.
**Habitat** burrowing in sand from extreme lower shore down to about 55m, so that its local distribution hardly overlaps *A. tenuis* (above).
**Distribution** Mediterranean, Atlantic, English Channel, North Sea and Baltic.

### *Arcopagia crassa* Pennant *( =Tellina crassa)*
**Blunt Tellin**
**Shell** up to 62.5mm long; oval; larger, heavier and plumper than either *A. tenuis* or *F. fabula* (above); valves almost equal, left not as convex as right; edge smooth; hinge with 2 cardinal teeth and 1 anterior lateral tooth and 1 posterior lateral tooth on each valve; umbones posterior from centre; external ligament; narrow lunule; posterior adductor scar shorter and thicker than anterior; outer surface with conspicuous concentric lines.
**Colour** periostracum ochre; interior red-yellow with white outer margin.
**Habitat** burrowing in mud, sand and gravel from 1–150m.
**Distribution** Atlantic, English Channel and North Sea.

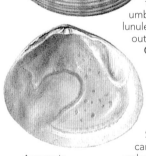

*Arcopagia
crassa*

### *Gastrana fragilis* (Linnaeus)
**Shell** up to 45mm long; valves similar; edge smooth; 2 cardinal teeth on both valves, lateral teeth lacking; umbones anterior of mid-line; posterior of shell extended and more pointed than anterior; conspicuous external ligament;

posterior adductor scar thicker than anterior; outer shell with irregular concentric lines.
**Colour** white to grey-yellow; periostracum pale brown.
**Habitat** burrowing in mud, clay and sand in shallow water, sometimes in estuaries.
**Distribution** Mediterranean and Atlantic north to Norway.

### *Macoma balthica* (Linnaeus)
**Baltic Tellin**
**Shell** up to 25mm long; similar to *A. tenuis* (opposite) but plumper and heavier; valves almost similar; edge, smooth; hinge with 2 small cardinal teeth but lacking lateral ones; umbones about central; external ligament; posterior adductor scar thicker than anterior one: outer shell with fine concentric lines.

*Gastrana fragilis*

*Macoma balthica*

**Colour** variable; exterior white-pink-purple and banded; interior pink-purple; periostracum often colourless.
**Habitat** burrowing in mud and sand in shallow brackish water, e.g. estuaries.
**Distribution** Atlantic north from Spain, English Channel, North Sea and Baltic.

*Scrobicularia plana*

### *Scrobicularia plana* (da Costa)
**Peppery Furrow Shell**
**Shell** up to 63mm long; light, oval and flattened; valves similar; hinge with 2 cardinal teeth on left valve and 1 on right; posterior adductor scar shorter than anterior; periostracum on periphery of valves; outer surface with concentric lines.
**Siphons** very long.
**Colour** periostracum brown; exterior grey-pale yellow; interior white.
**Habitat** burrowing in mud and sand between the tidemarks, often in salinities lower than *Macoma balthica* (above).
**Distribution** Mediterranean, Atlantic, English Channel, North Sea and Baltic.

### *Abra alba* (Wood)
**Shell** up to 20mm long, a little less deep, thin, outline oval; valves equal; anterior edge rounded, posterior edge slightly pointed; hinge with small teeth; umbones just posterior of midline.
**Colour** outside and inside white.
**Habitat** burrowing in mud, from shallow water down.
**Distribution** Mediterranean, Atlantic north to Norway, English Channel and North Sea.

*Abra alba*

*Gari
depressa*

### *Gari depressa* (Pennant)
**Large Sunset Shell**
**Shell** up to 6.3mm long; oval; valves almost equal; conspicuous posterior gape between valves; edge smooth; 2 diverging cardinal teeth on right valve, 1 diverging tooth and 1 simple cardinal tooth on left valve, no lateral teeth; umbones slightly forward; external ligament; stout periostracum; outer shell has rayed markings and with fine concentric lines and ridges.
**Siphons** 2; relatively short and separate.
**Colour** exterior pink; interior white-purple; periostracum green-brown.
**Habitat** burrowing in sand from extreme lower shore down to about 50m.
**Distribution** Mediterranean, Atlantic, English Channel and North Sea.

### *Pharus legumen* (Linnaeus)
**Shell** up to 125mm long; long and narrow; valves similar, long edges (dorsal and ventral) not parallel; beaks and umbones reduced and a little forward from the mid-line in front of conspicuous, black external ligament.
**Colour** exterior whitish with fine concentric lines; yellow-green-brown periostracum.

*Pharus
legumen*

**Habitat** burrowing in sand on extreme lower shore and in shallow water.
**Distribution** Mediterranean and Atlantic north to southwest Britain and Ireland.

### *Ensis ensis* (Linnaeus)
**Shell** up to 125mm long; long, narrow and curved; valves similar; beaks and umbones reduced and positioned at anterior; shell tapers towards posterior; dark external ligament.
**Siphons** short and united, except at the tip.

*Ensis
ensis*

**Colour** exterior whitish with red and brown patterns; yellow-green periostracum.

**Habitat** burrowing in sand on extreme lower shore and in shallow water.
**Distribution** Mediterranean, Atlantic, English Channel and North Sea.

### *Ensis siliqua* (Linnaeus)
**Pod Razor Shell**
**Shell** up to 200mm long; long, narrow; valves similar; dorsal and ventral edges almost parallel; beaks and umbones reduced and at the anterior; not tapering posteriorly; dark external ligament.
**Siphons** similar to *E. ensis* (above).
**Colour** exterior whitish and finely lined both vertically and horizontally, patterned with red; glossy yellow-green periostracum.

*Ensis
siliqua*

**Habitat** burrowing in sand on extreme lower shore and down to about 35m.
**Distribution** Mediterranean, Atlantic, English Channel and North Sea.

### *Solen marginatus* Montagu
**Grooved Razor Shell**

**Shell** up to 125mm long; narrow; similar to *E. siliqua* (opposite), but with a conspicuous, almost vertical groove or constriction just behind the anterior edge in front of the reduced beaks and umbones.

**Colour** exterior pale yellow with fine lines; pale brown periostracum.

**Habitat** burrowing in sand on the lower shore and in shallow water.

**Distribution** Mediterranean, Atlantic, English Channel and North Sea.

*Solen marginatus*

### *Mya truncata* Linnaeus
**Blunt Gaper**

**Shell** up to 75mm long; solid; left valve not as convex as right and bearing spoon-shaped process which projects towards the right valve and to which the internal ligament is attached; equivalent process in the right valve does not project; external ligament; conspicuous gape at posterior of shell as shown.

**Colour** dirty white-cream.

**Habitat** burrowing in mud and sand from middle shore down to 70m.

**Distribution** Atlantic, English Channel and North Sea.

right valve

*Mya
truncata*

posterior
view

left valve

### *Mya arenaria* Linnaeus
**Sand Gaper**

**Shell** up to 150mm long; more oval than *M. truncata* (above) but has a similar arrangement of valves and hinge processes.

**Colour** exterior dirty white-white; interior brown.

**Habitat** burrowing in mud and sand from the lower shore down to 70m, and in estuaries.

**Distribution** Atlantic, English Channel, North Sea and Baltic.

### *Hiatella arctica* (Linnaeus)

**Shell** up to 38mm long; irregularly shaped (frequently individuals are different shapes); edge smooth; hinge teeth usually worn away in adults; beaks and umbones well forward; external ligament; round anterior adductor scar slightly smaller than posterior one; shell with uneven concentric lines.

**Colour** exterior white-yellow; interior white; yellow-brown periostracum.

**Habitat** boring into soft rock or occupying holes already there, being anchored by byssus threads, on lower shore and in shallow water.

**Distribution** Mediterranean, Atlantic, English Channel and North Sea.

*Mya arenaria*
left valve

*Hiatella
arctica*

193

### *Gastrochaena dubia* (Pennant)
### Flask Shell

*Gastrochaena dubia*

**Shell** up to 25mm long; smooth; valves similar; hinge lacks teeth; beaks and umbones well forward; external ligament: posterior scar longer than anterior; conspicuous ventral gape at front (as illustrated).
**Siphons** united.
**Colour** exterior white-pale brown; interior white.
**Habitat** boring into soft rocks or firm sand on the extreme lower shore and in shallow water; a characteristic flask-like cavity is formed which is lined with particles of shell and secretion from the animal.
**Distribution** Mediterranean, Atlantic north to south-west Britain and Ireland.

ventral view

*Pholas dactylus*

### *Pholas dactylus* Linnaeus
### Common Piddock

**Shell** up to 150mm long; light; valves similar; edge smooth except anteriorly where it may be crenulate; umbones forward and reflected back on the shell; 4 accessory shell plates are situated dorsally; wide anterior ventral gape (as illustrated); periostracum covers the sculptured outer shell which bears concentric and radiating ribs which are roughest in front where they assist with drilling; interior of shell with 2 free teeth or apophyses (1 per valve); long, united siphons covered by horny sheath.
**Colour** pale yellow periostracum; exterior white-grey; interior white.

ventral view

**Habitat** boring into soft rock, wood, firm sand or peat on lower shore and in shallow water.
**Distribution** Mediterranean, Atlantic north to south-west Britain and Ireland and English Channel.

### *Zirfaea crispata* (Linnaeus)
### Oval Piddock

Shorter and stubbier than *Pholas dactylas* (above).
**Shell** up to 90mm long; valves similar; edge smooth apart from anterior; valves hardly meet, apart from the forward umbones, and at one point ventrally; outer shell with concentric ridges, rougher at front; inner shell with 1 strong tooth or hypophysis per valve.
**Siphons** long and united; covered in horny sheath.
**Colour** interior white; brown periostracum.
**Habitat** boring in clay and soft rock on lower shore and in shallow water.
**Distribution** Atlantic north from Biscay, English Channel and North Sea.

*Zirfaea crispata*

ventral view

### *Teredo navalis* Linnaeus
**Ship Worm**
**Shell** reduced; each valve has an inner tooth or hypophysis, shell functions as a drill and encloses only part of the worm-like animal; mantle secretes a hard, chalky tube up to 200mm long which follows the course of the animal as it bores through wood; a pair of special, hard pallets can close off the open end of the tube when the siphons have been withdrawn; the pallets are up to 5mm long.
**Colour** white.
**Habitat** boring in submerged wooden structures such as pilings and boats.
**Distribution** Mediterranean, Atlantic, English Channel, North Sea and west Baltic. N.B. several related species occur in the region.

pallets

tube

anterior

valves

*Teredo navalis*

### *Thracia papyracea* (Poli)
**Paper Thracia**
**Shell** up to 35mm long; fragile; left valve smaller than right; beaks and umbones back from mid-line; hinge lacks teeth; internal ligament and external ligament; outer shell surface with concentric lines.
**Colour** white.
**Habitat** on sand from the extreme lower shore downward.
**Distribution** Mediterranean, Atlantic, English Channel and North Sea.

*Thracia papyracea*

### *Pandora albida* (Röding)
**Pandora Shell**
**Shell** up to 38mm long; asymmetrical valves not equal; left valve trough-like, right valve flat; beaks and umbones well forward; internal ligament; adductor scars about equal.
**Colour** exterior white with fine concentric lines; pale brownish periostracum.
**Habitat** on sand and mud usually in shallow water.
**Distribution** Atlantic, English Channel and North Sea.

*Pandora albida*

# CLASS CEPHALOPODA Cuttlefishes, squids and octopuses

Uncoiled, cylinder- or sac-shaped molluscs. The foot is divided into a number of suckered tentacles surrounding the mouth and joining the head. The head bears conspicuous eyes and is immediately connected with the abdomen (visceral region). A shell is sometimes present internally, but rarely externally, and is sometimes lacking. The mouth leads to a pair of horny jaws resembling a parrot's beak; the radula is like a tongue. The sexes are separate. The animals are efficient predators and rapid movers. About 750 species occur in the world's seas and oceans.

## ORDER DECAPODA
### *CUTTLEFISHES AND SQUIDS*

cuttle bone
of *Sepia*

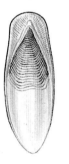

*Sepia officinalis*

Cephalopods with cylinder-shaped bodies bearing lateral fins. There is an internal shell or cuttlebone. The mouth is surrounded by 10 tentacles of which 2 are usually much longer than the remaining 8, and may be extended or retracted.

***Sepia officinalis*** (Linnaeus)
**Common Cuttlefish**
**Length** up to 300mm.
**Body** relatively broad and somewhat flattened so as to be oval in cross-section; conspicuous funnel on the underside near the head, paired fins run from behind the head to the tip of the body.
**Colour** very variable; may be black-brown striped or mottled above, paler to white below, but individuals are capable of rapid colour change, especially when threatened; the animal may also take the colour or patterning of its background.
**Habitat** over sand and in bays and estuaries, sometimes among eel-grass (*Zostera* sp. see p64–65).
**Distribution** Mediterranean, Atlantic, English Channel and North Sea. N.B. the cuttlebone is often washed ashore.

***Sepia elegans*** d'Orbigny
(Not illustrated)
**Length** up to 120mm.
**Body** similar to *S. officinalis* (above) but not so broad.
**Colour** red-brown above.
**Habitat** as for *S. officinalis*.
**Distribution** Mediterranean and Atlantic.

### *Sepiola atlantica* d'Orbigny

*Sepiola atlantica*

**Length** up to 50mm.
**Body** relatively short and cup-shaped, and bearing a pair of flap-like fins which do not run along its entire length.
**Colour** very variable; black-brown to pale above, usually pale below.
**Habitat** swimming over or burrowing in sand in shallow water.
**Distribution** Atlantic and English Channel.

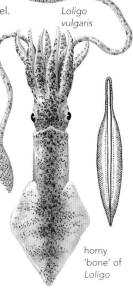
*Loligo vulgaris*

### *Loligo vulgaris* Lamarck
**Common Squid**
**Length** up to 500mm.
**Body** torpedo-shaped, with paired fins running about half way along, and joining together at the tip; viewed from above these fins have a diamond shaped appearance; a small, shield-like part of body projects slightly over the head; shell horny and somewhat pen-like. Suckers in the middle of the sucker group at the tips of the long tentacles are big, about four times the diameter of the ones at the margin.
**Colour** variable; often pink-white with purple-brown mottling above.
**Habitat** seldom found close to the shore.
**Distribution** Mediterranean, Atlantic, English Channel and occasionally in the North Sea and west Baltic.

### *Loligo forbesi* Seenstrup
**Long-finned Squid**
(Not illustrated)
**Length** up to 600mm.
**Body** similar to *L. vulgaris* (above).
**Colour** variable, but pink, red and brown predominate. Suckers in the middle of the sucker group at the tips of the long tentacles are not big, less than two times diameter of the ones at the margin.
**Habitat** seldom found close to the shore.
**Distribution** Atlantic, English Channel, North Sea and occasionally in the west Baltic.

horny 'bone' of *Loligo*

### *Alloteuthis subulata* (Lamarck)
**Length** up to 150mm.
**Body** similar to *Loligo* (above), but much smaller; paired narrow fins do not give a diamond-shaped appearance when viewed from above butare more lance-shaped; small, shield like part of body projects a little over the head.
**Colour** pale grey with brown-purple spots.
**Habitat** seldom found close to the shore.
**Distribution** Atlantic, English Channel and North Sea.

*Alloteuthis subulata*

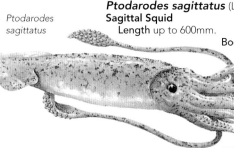

*Ptodarodes sagittatus*

### *Ptodarodes sagittatus* (Lamarck)
**Sagittal Squid**
**Length** up to 600mm.

**Body** torpedo-shaped, paired fins running about one-third of the way along and joining at the tip; no small shield-like part of the body projecting over part of the head; the 2 long tentacles are not retractile and can always be seen.

**Habitat** often found in surface waters at night.

**Distribution** Mediterranean, Atlantic, rarely in North Sea and west Baltic.

## ORDER OCTOPODA
## *OCTOPUSES AND THEIR ALLIES*

Cephalopods with bag-like bodies and no internal shell. An external shell is rarely present. There are no fins, although the 8 arms which carry suckers may be linked at their base by a web of skin.

### *Argonauta argo* Linnaeus
**Paper Nautilus**
**Length** adult females up to about 200mm; adult males up to 10mm.
**Body** female largely encased in a white, very thin, paper-like shell; 2 of the 8 arms are specialized for holding the shell so they fold back over the body and are usually appressed to the shell; male body dwarf, lacking a shell.
**Colour** highly variable; can change quickly according to situation, as can the form of patterning; ranges from silver-white- green-grey-red-blue with and without spots.

*Octopus vulgaris*

**Habitat** may creep on the bottom, but often found swimming.
**Distribution** Mediterranean and Atlantic.

### *Octopus vulgaris* Lamarck
**Common Octopus**
**Length** up to 1m overall but often no more than 600mm.
**Body** strong arms bear 2 rows of suckers; upper side of body often covered in warts.
**Colour** variable; grey-yellow-brown-green according to situation.
**Habitat** among rocks and stones and often in a lair where stones have been arranged for camouflage and protection.
**Distribution** Mediterranean, Atlantic north to English Channel.

### *Eledone cirrhosa* (Lamarck)
**Lesser Octopus**
**Length** up to 500mm.
**Body** arms bear 1 row of suckers; may be smooth or warty.
**Colour** predominently red-brown above and white below.
**Habitat** among rocks and stones, occasionally at extreme lower shore.
**Distribution** Atlantic, English Channel and northern North Sea.

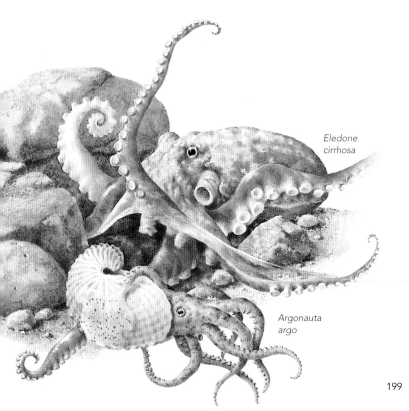

*Eledone cirrhosa*

*Argonauta argo*

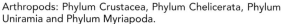

▼▶ Planktonic
crustacean larvae

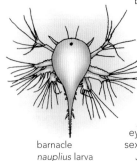

barnacle
*nauplius* larva

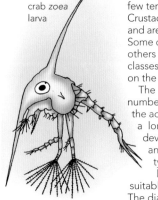

crab
*megalopa*
larva

crab *zoea*
larva

**Arthropods: Phylum Crustacea, Phylum Chelicerata, Phylum Uniramia and Phylum Myriapoda.**

These bilaterally symmetrical animals have segmented bodies which have developed from three embryonic cell layers. In most cases there is a reduced body cavity (true coelom), and a large blood space. Usually each body segment carries a pair of jointed appendages, one of which forms the jaws. In some parts of the body the segments may fuse (e.g. the head) and the precise relationship of the appendages to the segments may not be clear. In the crustacea the exoskeleton is composed of calcium carbonate and other materials whilst in the insects it is formed from chitin and other substances. It is hard and jointed. A well developed nervous system is present with eyes and other sense organs. True nephridia are lacking. The sexes are usually separate.

Jointed-limbed animals include the marine and freshwater crustaceans, the familiar insects (the commonest animals in the world) and the centipedes and spiders. Until fairly recently the features listed above led to the arthropods being considered as one phylum. However recent research on their functional morphology and developmental biology has led to their being considered as a number of separate phyla as shown here.

## Phylum Crustacea

The crustaceans form the major group of marine arthropods and this phylum is well represented in most marine habitats. A much smaller but nevertheless interesting marine group is the Pycnogonida (the sea spiders) which are classified along with the terrestrial spiders and scorpions in the Phylum Chelicerata. Members of the Phylum Crustacea have a body divided into three parts – head, thorax and abdomen. The exoskeleton is generally supported by calcium salts and is often heavy. The head bears two pairs of antennae (one pair on each of the second and third segments), and mandibles (jaws) on the fourth segment. Although the phylum is almost entirely aquatic, there are a few terrestrial exceptions, for instance, the familiar woodlouse. Crustaceans show a diverse range of body forms and life styles, and are divided up into several classes and a number of orders. Some of these comprise of microscopic, planktonic organisms, others are found solely in fresh water. Representatives of two classes (Cirripedia and Malacostraca) are commonly met with on the shore and in the shallow sea.

The basic crustacean body plan has been modified in a number of ways. In the Cirripedia (barnacles and their allies) the adults do not closely resemble other crustaceans, and for a long time they were classified elsewhere. Their larval development follows the typical crustacean form, however, and a small, free-swimming organism passes through two typically crustacean stages (nauplius and cypris) before leaving the plankton to settle and metamorphose on suitable rock or shell where it grows into an adult barnacle. The diagrams pictured here show a number of different types of crustacean larvae which may be taken from the plankton at particular times of the year. Whilst they are clearly crustaceans, these larvae do not closely resemble the adult form into which they will grow. Crustaceans are found in almost all types of marine environment. The barnacles have evolved to lead a sessile life as adults, with their bodies protected by a number of massive plates of calcium carbonate which form the 'shell'. The type of shell base which attaches them to the rocks may help with identification; it may be calcareous or membranous, and can be seen if a specimen is removed from the rock.

barnacle
*cypris* larva

Effectively they are attached to the rocks by their head ends, and their thoracic appendages have become modified as filter-feeding organs. These can be projected outside the 'shell' and swept backwards and forwards through the water to collect any appropriate food particles.

The Malacostraca is a large and important class. It includes a great variety of organisms whose basic body plan is indicated below. Many malacostracan

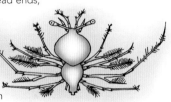

crayfish
*phyllosoma* larva

appendages are biramous, i.e. they branch in two. Where this is the case one branch of the limb may have developed to carry out one function – for instance walking – and the other a different function – for instance respiration. This particular situation occurs in the thoracic legs of many malacostracans, where the shorter branch of the limb is modified to form a gill.

Members of the Malacostraca are generally free living and actively seek out their food using their well-developed eyes and chemoreceptors. This is especially true of many members of the order Decapoda such as crabs and lobsters, which often lead a scavenging or predatory life on the seabed. (They are arranged as a suborder Reptantia = creepers.) The prawns and shrimps also belong to this order, but their way of life is different in that they can swim as well as walk. Indeed, some members of this group (arranged as a suborder Natantia = swimmers) live entirely by swimming in the waters of the ocean, and never touch the seabed.

A swimming life is generally the rule for the malacostracan larvae of all groups. After fertilization, the female may carry the developing embryos attached to the outside of her body in a manner that allows her to keep them well aerated. When the larvae emerge from the egg cases they find their way to the surface waters and become members of the plankton where they feed and grow. During this period they may change their form once or more before they are finally ready for metamorphosis, and the juveniles develop to begin life as adults on the seabed or elsewhere. For more information see Newell, G. E. and Newell, R. C. 1973.

▼ Basic plan of the malacostracan body

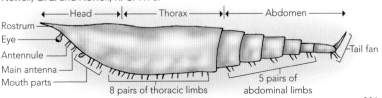

Head ── Thorax ── Abdomen

Rostrum
Eye
Antennule
Main antenna
Mouth parts
8 pairs of thoracic limbs
5 pairs of abdominal limbs
Tail fan

## CLASS CIRRIPEDIA Barnacles

Crustaceans which as adults have a calcified exoskeleton comprised of several chalky plates, forming a shell. For identification of other barnacles see Darwin, C. 1851–1854 and O.E.C.D. catalogues on barnacles. Rainbow, P. S. 1984 gives details of British barnacle biology.

### *Lepas anatifera* (Linnaeus)
**Goose Barnacle**
**Shell** about 50mm long with 5 translucent, nearly white plates which have a bluish-grey tinge.
**Stalk** 100–200mm long, somewhat retractable, and with a brown-grey skin.
**Habitat** pelagic; normally attached to boats and driftwood.
**Distribution** Atlantic, English Channel and North Sea.

Lepas
anatifera

Dosima
fascicularis

Scalpellum
scalpellum

### *Dosima fascicularis* (Ellis & Solander)
**Buoy-making Barnacle**
**Shell** about 35mm long with 5 translucent, brittle, white plates.
**Stalk** secretes a spongy white float.
**Habitat** pelagic; often washed ashore after south-westerly gales.
**Distribution** Atlantic and English Channel.

### *Scalpellum scalpellum* (Linnaeus)
**Shell** about 20mm long with 14 small, hairy, grey-white plates.
**Stalk** is covered with small scales and nearly as long as shell.
**Habitat** attached to hydroids, bryozoans and worm tubes, usually in deeper water.

group
growing on
dead hydroid

individual

**Distribution** Mediterranean, Atlantic, English Channel and North Sea. Similar species *Pollicipes pollicipes* (Gmelin) (Not illustrated). Stalk much shorter than shell, not more than 50mm overall.

### *Verruca stroemia* (O. F. Müller)
Sessile.
**Shell** asymmetrical, flattened, up to 5mm in diameter with 4 unequal, ribbed plates which may be grey, white or brown; base membranous if detached.
**Habitat** under stones and attached to shells from lower shore down to about 70m.
**Distribution** Mediterranean, Atlantic, English Channel and North Sea. N.B. may be quite flat.

*Verruca stroemia*

### *Chthamalus montagui* (Southward)
Sessile.
**Shell** slightly asymmetrical, up to 12mm in diameter with 6 grooved plates arranged as a cone; lateral plates overlap terminal plates so that these appear narrow (in older specimens the fusion of plates may make their delineation difficult); kite-shaped opening; base membranous if detached.
**Colour** usually whitish.
**Habitat** on rocks on upper and middle shore; range dependent on exposure always found above *Semibalanus balanoides* (see below) on the shore.
**Distribution** Mediterranean, Atlantic and English Channel. Another species *C. stellatus* has an oval opening.

*Chthamalus montagui*

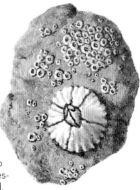

### *Semibalanus balanoides* (Linnaeus)
**Acorn Barnacle**
Sessile.
**Shell** symmetrical, up to 15mm in diameter with 6 plates; shape flat or steeply conical; lateral plates overlap one terminal plate, but are themselves overlapped by the other; 1 terminal plate appears narrow and the other wide, diamond-shaped opening, base membranous if detached.
**Colour** usually dirty white.
**Habitat** on rocks; in northern areas where *Chthamalus* species (above) are absent from high water mark for neap tides downward; in southern areas where *Chthalamus* is present, below this species from about middle shore downward.
**Distribution** Atlantic, English Channel, North Sea and west Baltic.

*Semibalanus balanoides*

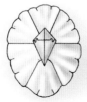

Form of shell plates in *C. montagui*

Form of shell plates in *S. balanoides*

### *Balanus crenatus* Bruguière
Sessile.

**Shell** conical, up to 20mm in diameter and nearly as tall as it is round; body leans over when viewed sideways; 6 white-grey plates; base calcareous if detached; rim of inside movable plates yellow and purple.

**Habitat** on the lower shore and in shallow water; not on exposed rock surfaces.

**Distribution** Atlantic, English Channel, North Sea and the west Baltic.

*Balanus crenatus*
details of opening

### *Balanus balanus* (Linnaeus)
(Not illustrated) Sessile.

**Shell** large, conical, up to 30mm in diameter with 6 stout, prominently ribbed plates; opening roughly triangular; base calcareous if detached.

**Colour** whitish.

**Habitat** on lower shore down to deep water on rocks and shells, e.g. *Pecten* (see page 179).

**Distribution** Atlantic north from English Channel, North Sea and west Baltic.

### *Balanus perforatus* Bruguière
Sessile.

**Shell** fairly symmetrical, up to 30mm in diameter, tall and conical with 6 grey-purple-brown smooth or lined plates, often separated at the apex leaving a jagged lip; opening of shell off centre; base calcareous if detached.

**Habitat** lower shore where it is not as numerous as *S. balanoides* (page 203) and occupies the zone just below it.

*Balanus perforatus*

**Distribution** Mediterranean and Atlantic north to the English Channel.

### *Balanus improvisus* Darwin
Similar to *B. crenatus* (above). Sessile.

**Shell** up to 15mm in diameter, symmetrical rim of inside movable plates white with purple bands; base calcareous if detached.

**Habitat** in brackish water such as estuaries and lagoons.

*Balanus improvisus*
detail of opening

**Distribution** Atlantic, English Channel and North Sea regions; it is the only barnacle present in the Baltic Sea proper.

### *Acasta spongites* Darwin
Sessile.

**Shell** up to 12mm in diameter; easy to identify as it is always partially or totally embedded in a sponge and because its base is not flat, but convex.

**Habitat** associated with a sponge (usually *Dysidea fragilis*, see page 75) on extreme lower shore also and downward.

*Acasta spongites* in
*Dysidea fragilis*

**Distribution** Mediterranean, Atlantic and English Channel.

### Megatrema anglicum (Leach)

(Illustrated attached to *Caryophyllia smithii* on page 100)
**Shell** up to 3mm in diameter; easy to identify because it is always attached to a coral.
**Colour** grey-green-brown.
**Habitat** as for the coral; usually found on *Caryophyllia smithii*, but occasionally on other genera including *Cladocora*, *Denodrophyllia* and *Balanophyllia*.
**Distribution** Mediterranean, Atlantic, and English Channel.

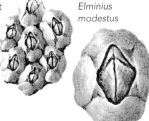

*Elminius modestus*

### Elminius modestus Darwin

Sessile.
**Shell** flattened, up to 10mm in diameter with 4 usually distinct plates in the shell; plates smooth with 2 slight depressions.
**Colour** white-grey.
**Habitat** attached to rocks on upper middle shore, sometimes near the influence of fresh water.
**Distribution** this Australian barnacle first appeared near Southampton in the 1940s and is spreading rapidly in the Atlantic, English Channel and North Sea areas.

*Conchoderma virgatum*

### Conchoderma virgatum (Spengler)

**Shell** 5 small skeletal plates on thick skin; overall height up to 50mm; purplish-red to dark red-brown with stripes, leathery.
**Stalk** widening from attachment point to main body.
**Habitat** on various mobile objects including the bottom of boats and ships.
**Similar species** *C. patula* (Ranzani) (not illustrated) which has ear-like tubes on the extremity of the body.
**Distribution** Mediterranean and Atlantic north to west of UK and Ireland.

### Sacculina carcini Thompson
**Parasitic Barnacle**

Parasitic and quite unlike the other barnacles; appears as a conspicuous lump covered with pale yellow-brown skin attached under the abdomen of the crab *Carcinus maenas* (see page 237), and sometimes to other species, in such a way as to prevent the abdomen from folding closely under the cara-pace; readily distinguished from the crab's own eggs which may be brooded in this position by the female, because the egg mass is granular in texture and *S. carcini* is smooth.
**Habitat** and **Distribution** as for the host.

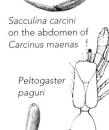

*Sacculina carcini*
on the abdomen of
*Carcinus maenas*

### Peltogaster paguri Rathke

A parasite, like *Sacculina carcini* (above), but appearing as a smooth elongated growth on the side of the abdomen of hermit crabs; not visible unless the hermit crab is remove from its gastropod shell; found on *Pagurus bernhardus* and *P. cuanensis*.
**Habitat** as for the host.
**Distribution** Mediterranean, Atlantic, English Channel and North Sea.

*Peltogaster
paguri*

# CLASS MALACOSTRACA

Crustaceans with compound eyes, 2 pairs of antennae, and with the head and thorax fused and usually covered by a carapace. The thorax has 8 segments each bearing appendages (first 3 pairs assist with feeding, and the remaining 5 pairs usually develop as legs). The abdomen usually comprises of 6 segments, with appendages variously modified for swimming, reproduction, etc.

## ORDER LEPTOSTRACA

Small malacostracans with stalked eyes separated by a rostral spine. The carapace enfolds all the thoracic and 4 pairs of abdominal appendages.

### Nebalia bipes (Fabricius)
**Length** up to 10mm. Uppermost pair of antennae short, lowermost longer.
**Habitat** middle and lower shore.
**Distribution** Mediterranean, Atlantic, English Channel and North Sea.

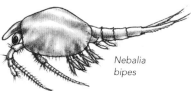

*Nebalia bipes*

## ORDER STOMATOPODA

Moderately sized malacostracans possessing a short, shield-shaped carapace. The last joint of the 2nd pair of thoracic legs forms a spiny crusher.

### Squilla desmaresti Risso
**Length** up to 120mm. Last joint of 2nd pair of thoracic legs bears 4 conspicuous spines and terminates in a sharp point.
**Habitat** in sand and mud from extreme lower shore down to deeper water.
**Distribution** Mediterranean, Atlantic and English Channel.

*Squilla desmaresti*

# ORDER CUMACEA

Small malacostracans with bodies compressed sideways. The distinctively shaped carapace covers only part of the thorax. For identitication of other cumaceans see Jones, N. S. 1976.

### *Pseudocuma longicornis* (Bate)
**Length** up to 5mm. Anterior part of carapace appears ridged.
**Habitat** burrowing in sand or mud on lower shore and in shallow water; some-times swarming in surface waters.
**Distribution** Mediterranean, Atlantic, English Channel and North Sea.

*Pseudocuma longicornis*

# ORDER TANAIDACEA

Small malacostracans with bodies compressed dorso-ventrally. The reduced carapace covers only the first 2 segments of the thorax; 2nd pair of thoracic appendages bears small pincers and these are followed by 6 pairs of similar legs. Abdominal appendages bear swimmerets, but a tail fan is lacking.

### *Apseudes latreillei* (Milne-Edwards)
**Length** up to 6mm. Antennae branched; terminal appendages branched.
**Habitat** often under stones or in mud among seaweeds on lower shore and in shallow water.
**Distribution** Mediterranean, Atlantic, English Channel and North Sea.

*Apseudes latreillei*

# ORDER EUPHAUSIACEA

Small, pelagic malacostracans with conspicuous eyes. The carapace bears a rostrum covering the thorax, but not shielding the thoracic appendages. The last 7 pairs of thoracic appendages are clearly branched near their points of origin; frequently they bear no pincers. Allen, J. A. 1967 gives a fuller account of the euphausiaceans.

### *Meganyctiphanes norvegica* (M. Sars)
**Krill**
**Length** up to 15mm. Carapace bears rostrum, being about one-third of the body length; lower margins are smooth and not toothed nor horned as in some species; characteristic outgrowths on either side of the head behind the eyes.
**Habitat** pelagic.
**Distribution** Mediterranean, Atlantic, English Channel and North Sea.

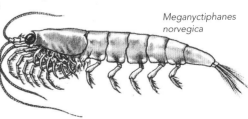

*Meganyctiphanes norvegica*

## ORDER MYSIDACEA OPOSSUM SHRIMPS

Small, swimming malacostracans which seldom reach more than 30mm in length. They possess a thin carapace and a long abdomen of 6 segments. The branched, thoracic appendages lack pincers, and the abdominal appendages are short. A tail fan is borne on the terminal segment, the precise form of which is useful in identification, and which is shown separately where the illustration is a side view. Conspicuous eyes are carried on movable eyestalks. These animals are often transparent and may remain suspended in the water by their swimming movements; females often carry a brood pouch on their under surfaces as the illustration of *Anchialina* shows. For fuller details of many European mysids see Tattersall, W. M. and Tattersall, O. F. 1951.

### Anchialina agilis (G. O. Sars)

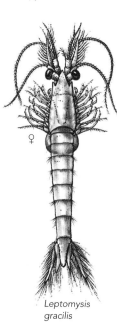

Anchialina
agilis
♀

tail
detail

**Length** up to 9mm. Carapace relatively large, covering the whole thoracic region and reaching more than half the length of the abdomen; outer body surface covered with minute bristles or spines, especially noticeable on the terminal segment.
**Habitat** on the seabed by day from shallow water down to 60m or deeper; migrating quickly to the surface for a relatively short period by night.
**Distribution** Mediterranean, Atlantic, English Channel and southern North Sea.

### Leptomysis gracilis (G. O. Sars)

**Length** up to 13mm. Carapace short, does not cover the whole thoracic region, and is very little wider than the abdomen; entire body surface covered with minute scales (visible with a strong hand lens).
**Colour** transparent and nearly colourless; abdominal region may have slight yellow-red tinge.
**Habitat** lower shore in pools and on the seabed in shallow water; may be taken with plankton at certain times of the year.
**Distribution** Mediterranean, Atlantic, English Channel and North Sea.

### Leptomysis mediterranea G. O. Sars

(Not illustrated)
**Length** up to 10mm. Very similar to *L. gracilis* (above), but fine scales are totally absent on the opaque brown body surface.
**Habitat** sometimes in great numbers from lower shore down to 100m.
**Distribution** Mediterranean, Atlantic, English Channel and North Sea as far north as Scotland.

### Mysis relicta Louvén

**Length** up to 20mm. Carapace covers all but dorsal aspect of the last 2 thoracic segments and is about half the length of the slender abdomen; abdomen bent downwards in a characteristic posture giving a humped appearance.

♀

Leptomysis
gracilis

Habitat in fresh and brackish water, often in lakes and lagoons.
Distribution Venice, Gulfs of Bothnia and Finland and in the Baltic generally north to 58°N.

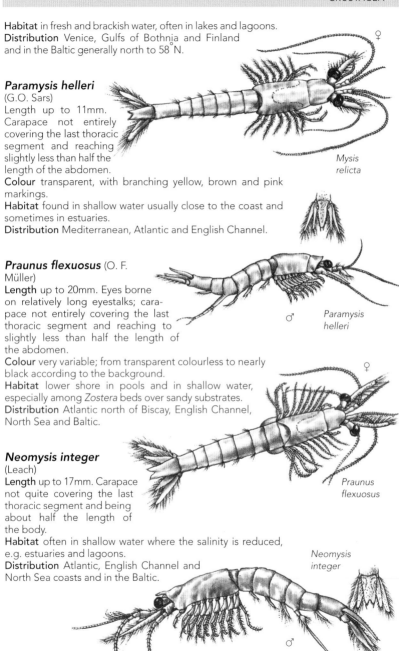

## Paramysis helleri
(G.O. Sars)
Length up to 11mm. Carapace not entirely covering the last thoracic segment and reaching slightly less than half the length of the abdomen.
Colour transparent, with branching yellow, brown and pink markings.
Habitat found in shallow water usually close to the coast and sometimes in estuaries.
Distribution Mediterranean, Atlantic and English Channel.

## Praunus flexuosus (O. F. Müller)
Length up to 20mm. Eyes borne on relatively long eyestalks; carapace not entirely covering the last thoracic segment and reaching to slightly less than half the length of the abdomen.
Colour very variable; from transparent colourless to nearly black according to the background.
Habitat lower shore in pools and in shallow water, especially among Zostera beds over sandy substrates.
Distribution Atlantic north of Biscay, English Channel, North Sea and Baltic.

## Neomysis integer
(Leach)
Length up to 17mm. Carapace not quite covering the last thoracic segment and being about half the length of the body.
Habitat often in shallow water where the salinity is reduced, e.g. estuaries and lagoons.
Distribution Atlantic, English Channel and North Sea coasts and in the Baltic.

*Mysis relicta*

*Paramysis helleri*

*Praunus flexuosus*

*Neomysis integer*

209

## ORDER ISOPODA

Resembling woodlice, isopods are small, dorso-ventrally flat-tened malacostracans lacking a carapace. The inner antennae are small, the outer antennae conspicuous, and the eyes are not stalked. The first pair of thoracic appendages are used as mouth parts, the remainder for locomotion. The first five pairs of abdominal appendages are leaf-like, functioning as gills, and the 6th segment bears branching swimming appendages which may combine with the telson. Naylor, E. 1972 provides a fuller account of many European isopods.

*Gnathia maxillaris*

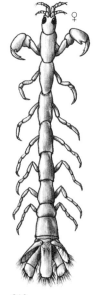

♀

♂ head of

### Gnathia maxillaris (Montagu)
Length up to 6mm; inner antennae longer than outer; sexes very different. Male head has large hooked jaws between antennae; viewed from above head has a shallow recess between jaws with a slight forward process in midline; body widest bewteen eyes; head and thorax wide, abdomen minute and tapering. Female head small, lacks jaws; thorax bulbous, widest in mid-thorax; abdomen small.
Habitat in crevices, dead barnacles and algal holdfasts.
Distribution Mediterranean and Atlantic north to south-west and southern Britain, English Channel.

head of ♂

*Anthura gracilis*

♀

### Anthura gracilis (Montagu)
Length female up to 11mm; male up to 4mm.
Head eyes relatively large; inner antennae short; outer antennae short in female and long and hairy in the male.
Body long; thin thorax; terminal joint of the 1st leg folded back to form a pincer against the next joint; not so in the remaining legs; abdomen relatively short and fan-like.
Habitat in crevices, worm tubes and among seaweeds; in shallow water.
Distribution Mediterranean, Atlantic and west English Channel.

### Limnoria lignorum (Rathke)
Length up to 4mm.
Head antennae short, together not wider than the abdomen.
Body thorax slightly longer and wider than abdomen which has almost parallel edges; 6th pair of abdominal appendages small and just extending beyond edge of telson.
Habitat boring in wood.
Distribution Atlantic from English Channel north, North Sea and west Baltic.

*Limnoria lignorum*

### Limnoria tripunctata (Menzies)
Length up to 4mm.
Head both pairs of antennae short.
Body thorax slightly longer and wider than abdomen, which has almost parallel edges; 6th pair of

abdominal appendages extending just beyond edge of telson; rear part of telson carries 3 small tubercles (visible with a lens).
**Habitat** often found near power station discharges boring in wood.
**Distribution** Mediterranean, Atlantic, English Channel and North Sea.

*Limnoria tripunctata*

### Eurydice affinis Hansen
**Length** up to 6mm.
**Head** inner antennae short, outer nearly as long as head and thorax together.

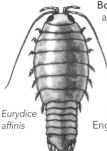

*Eurydice affinis*

**Body** thorax oval, abdomen not as wide and about two-thirds as long; 6th pair of abdominal appendages short and not extending much beyond edge of telson; rear of telson convex (visible with a lens); small black spots on dorsal surface only.
**Habitat** in sand on shore and in shallow water.
**Distribution** Mediterranean, Atlantic, English Channel and North Sea.

### Eurydice pulchra Leach
**Length** female up to 6.5mm; male to 8mm.
**Head** outer antennae nearly as long as both the head and thorax together.
**Body** thorax oval, wider than abdomen, and about one and a half times as long; 6th pair of abdominal appendages short and not extending much beyond the edge of telson.
**Habitat** in sand from middle shore downward.
**Distribution** Atlantic, English Channel, North Sea and the west Baltic.

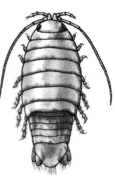

*Eurydice pulchra*

### Dynamene bidentata (Adams)
Sexes different.
**Male** up to 7mm.
**Head** rounded, convex, with conspicuous eyes.
**Body** 6th thoracic segment has one pair of small lateral processes with project back over 7th segment and 1st two abdominal segments; telson rough (visible with a lens); 6th pair of abdominal appendages with conspicuous outward-pointing terminal branch.
**Female** up to 6mm.
**Body** 6th thoracic segment lacks backward projections.
**Habitat** rock crevices and empty barnacle shells; juveniles among algae on shore.
**Distribution** Mediterranean and Atlantic north to English Channel and south-west Britain.

*Dynamene bidentata* ♂

Sphaeroma
serratum

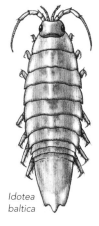

Idotea
baltica

Idotea
linearis

### Sphaeroma serratum (Fabricius)
Length female up to 10mm; male up to 12mm.
Head inner antennae about half the length of outer antennae; outer antennae about one-third of body length.
Body oval; 6th pair of abdominal appendages terminate in conspicuous, oval, flat joints with serrated outer edges; telson has a smooth upper surface.
Habitat in crevices and under stones, usually on middle shore.
Distribution Mediterranean, Atlantic north to British isles, and English Channel.

### Lekanesphaera rugicauda (Leach)
### ( =Sphaeroma rugicauda)
(Tail only illustrated) Similar to *S. serratum* (above).
Length female up to 7.5mm; male up to 10mm.
Body outer edge of 6th abdominal appendages not serrated; upper surface of telson covered with tubercles.
Habitat burrowing or under stones, etc., in estuaries and salt marsh pools.
Distribution Atlantic, English Channel, North Sea and Baltic.

tail

Lekanesphaera
rugicauda

### Idotea baltica (Pallas)
Length female up to 17mm; male up to 30mm.
Head inner antennae short; outer antennae about one-quarter body length.
Body oblong with slightly convex sides; abdomen narrower than thorax, terminating in long telson with almost straight edges and keeled upper surface; rear of telson produced into 3 teeth.
Habitat on seaweeds on lower shore and in shallow water.
Distribution Mediterranean, Atlantic, English Channel, North Sea and Baltic.

### Idotea granulosa Rathke
Similar to *I. baltica* (above).
Length female up to 13mm; male up to 20mm.
Body telson narrows sharply at first and ends in a conspicuous central spine; no keel.
Habitat on the shore among seaweeds.
Distribution Atlantic north from English Channel, North Sea and Baltic.

Idotea
granulosa

### Idotea linearis (Linnaeus)
Length male 40mm, female less.
Body inner antennae reach just beyond start of 2nd joint of outer antennae, which themselves reach about half-way along the slender body; rear edge of telson concave.
Habitat in shallow water near shore.
Distribution Mediterranean, Atlantic north to English Channel and south-west Britain.

### Idotea chelipes (Pallas) ( =I. viridis)
Similar to *I. baltica* (opposite).
**Length** female up to 10mm; male up to 15mm.
**Body** telson slightly keeled and with one tooth at the tip, this is not as well developed as in *I. granulosa* (opposite).
**Habitat** in brackish water.
**Distribution** Mediterranean, Atlantic, English Channel, North Sea and Baltic.

### Astacilla longicornis (Sowerby) ( =A. intermedia)
**Length** female about 30mm; male about 10mm.
**Head** outer antennae long and conspicuous.
**Body** slender and round; 4th thoracic segment much longer

*Astacilla longicornis*

than the preceding ones and separating the 1st 4 pairs of thoracic limbs from the shorter, relatively hairless remaining 3 pairs which clasp the spines of the sea-urchin *Echinus esculentus* (see page 263) upon which this isopod lives as a commensal; abdomen terminates in a point.
**Distribution** Atlantic, English Channel and North Sea.

*Idotea chelipes*

### Jaera albifrons Leach
**Length** female up to 5mm; male up to 2.4mm.
**Head** eyes quite large; inner antennae very short; outer antennae about half the length of the body.
**Body** roughly oval in outline; sometimes deep notches between segments of thorax whose edges are fringed with spines; 6th abdominal appendages very short, set in a notch and just projecting from under the telson.
**Habitat** on the upper shore under stones, and in estuaries.
**Distribution** Atlantic, English Channel, North Sea and Baltic.

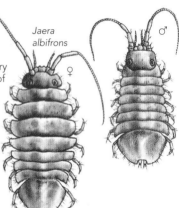

*Jaera albifrons*

### Jaera nordmanni (Rathke)
**Length** male up to 4.5mm, female up to 3.5mm.
**Head** with small eyes; inner antennae very short, outer entennae about half body length.
**Body** ovoid and flat with edges very hairy; deep notches apparent between the rear thoracic segments; abdomen rounded, with terminal appendages set in a shallow notch.
**Habitat** usually under stones on shore, especially in freshwater streams.
**Distribution** Mediterranean, Atlantic north to Scotland, English Channel.

*Jaera nordmanni*

213

**Ligia oceanica** (Linnaeus)
**Sea-slater**
*L. oceanica* strongly resembles a woodlouse.
**Length** up to 25mm.
**Head** outer antennae is about two-thirds of the body length.
**Body** flat, oval; 6th abdominal appendages long and trailing behind the telson.
**Habitat** on rocks above the intertidal zone.
**Distribution** Mediterranean, Atlantic, English Channel and North Sea.

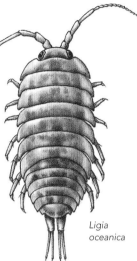

*Ligia oceanica*

# ORDER AMPHIPODA

Amphipods are small malacostracans lacking a carapace which are flattened laterally. The upper and lower antennae are variously developed, and the eyes are not stalked. The first pair of thoracic appendages are usually modified as mouth parts, the remaining six pairs being variously developed with pincers or claws for gripping. The abdomen bears three jumping legs, three swimming appendages and a telson, but sometimes it may be degenerate. Thorax and abdominal segments are less easy to distinguish than in isopods. Lincoln R. J. 1978 and Moore, P. G. 1984 provide a fuller account of the European amphipods

*Lysianassa ceratina*

**Lysianassa ceratina** (Walker)
**Length** up to 10mm.
**Head** upper antennae mounted on a thick base; lower antennae not much longer than upper in the female but nearly twice as long in the male.
**Body** 1st thoracic limb with very slender short pointed tip, 2nd slightly broader, blunt, flat.
**Habitat** among algae in shallow water.
**Distribution** Mediterranean, Atlantic north to Norway, English Channel and North Sea.

*Orchestia gammarella*

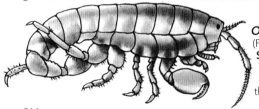

**Orchestia gammarella**
(Pallas)
**Sand-hopper**
**Length** up to 20mm.
**Head** upper antennae shorter than the large-jointed section of

the lower antennae and lacking any branch; lower antennae terminate in a number of small joints which appear smooth under a hand lens.
**Body** 2nd thoracic leg terminates in a small claw which acts as a minute pincer against the next joint; 3rd thoracic leg terminates in a larger pincer, the penultimate joint being large, oval and relatively conspicuous.
**Habitat** among stones and seaweeds, from upper to middle shore.
**Distribution** Mediterranean, Atlantic, English Channel, North Sea and Baltic.

### *Leucothoë spinicarpa* (Abildgaard)
**Length** up to 18mm.
**Head** lower antennae slightly shorter than upper antennae.
**Body** first thoracic limb formed into pincer using the terminal (6th) joint closing against a fixed projection of the 4th joint; terminal joint itself about one-third to half length of 5th joint; 2nd thoracic limb much larger with terminal (6th) joint closing to make a pincer against side of the expanded subterminal (5th) joint; telson 3 times as long as it is wide.
**Habitat** from shore down to 600m among algae, sponges and sea-squirts.
**Distribution** Mediterranean, Atlantic north to Ireland and Scotland, English Channel and North Sea.

*Leucothoë spinicarpa*

### *Talitrus saltator* (Montagu)
**Sand-hopper**
**Length** up to 16mm.
**Head** upper antennae shorter than the large-jointed section of the lower antennae and lacking any branch; lower antennae terminate in a number of small joints which appear rough and toothed under a hand lens.
**Body** 2nd thoracic leg terminates in a claw-like joint.
**Habitat** upper shore associated with rotting seaweeds.
**Distribution** Mediterranean, Atlantic, English Channel, North Sea and Baltic.

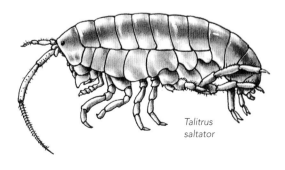

*Talitrus saltator*

### Gammarus locusta (Linnaeus)
**Length** female up to 14mm; male up to 20mm.
**Head** upper antennae have a small branch which is visible under water, lower antennae are shorter.

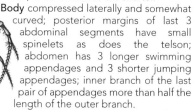

**Body** compressed laterally and somewhat curved; posterior margins of last 3 abdominal segments have small spinelets as does the telson; abdomen has 3 longer swimming appendages and 3 shorter jumping appendages; inner branch of the last pair of appendages more than half the length of the outer branch.
**Habitat** abundant under stones, etc., on middle to lower shores.
**Distribution** Mediterranean, Atlantic, English Channel, North Sea and Baltic.

*Gammarus locusta*

### Gammarus zaddachi Sexton; G. chevreuxi Sexton; G. duebeni Liljeborg
(Not illustrated) 3 species which closely resemble *G. locusta* (above). Rasmussen 1973 and Lincoln 1979 provides notes for distinguishing them. Generally they occur in brackish conditions.

### Elasmopus rapax da Costa
**Length** up to 10mm.
**Head** with upper antennae nearly half as long as body; lower antennae shorter.
**Body** 1st thoracic limb terminates in a fine pincer-like claw, 2nd similar but much larger; both quite hairy on posterior edges; 5th, 6th and 7th thoracic appendages with stout bases; remaining parts of limbs relatively slender.
**Habitat** from the shore down to 100m, often among algae.
**Distribution** Mediterranean, Atlantic north to Norway, English Channel and North Sea.

*Elasmopus rapax*

### Haustorius arenarius (Slabber)
**Length** up to 10mm.
**Head** eyes difficult to distinguish; upper antennae with a small branch that can be distinguished under water, and which are slightly shorter than the lower antennae, the basal joints of which are flattened; both pairs of antennae appear suspended from the hood-like head segment.

*Haustorius arenarius*

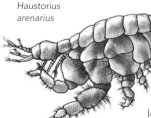

**Body** side-plates of body become larger towards the rear; last 3 pairs of thoracic legs with flattened plate-like joints.
**Habitat** burrowing in sand on the middle and lower shore.
**Distribution** Atlantic, English Channel and North Sea.

### Bathyporeia pelagica (Bate)
**Length** up to 6mm.
**Head** upper antennae appear to be borne on an outgrowth of the head, being short with a small branch which can be distinguished under water; lower antennae as long as the body in the male and twice as long as the upper antennae in the female.
**Habitat** lower middle shore, often burrowing in sand.
**Distribution** Atlantic, English Channel, North Sea and west Baltic.

*Bathyporeia pelagica*

### Tritaeta gibbosa (Bate)
**Length** up to 6mm. (Head detail only shown.)
**Head** with roughly equal upper and lower antennae in the female, upper a little shorter than lower in the male.
**Body** 1st and 2nd thoracic limbs with terminal joints acting as fine pincers against subterminal joints.
**Habitat** on the shore and down to 150m, often associated with sponges and sea-squirts.
**Distribution** Mediterranean, Atlantic north to Shetland, English Channel and North Sea. Does not occur on the shore in northern regions.

*Tritaeta gibbosa*

### Aora typica Kröyer
**Length** up to 9mm.
**Head** upper antennae bear a short side branch visible with a lens when animal is immersed; lower antennae shorter and more robust.
**Body** (male) in 1st thoracic limb the 3rd joint is extended into a large pointed process forming a spine which reaches the subterminal joint; terminal joint claw-like; female lacks these features. Body of both sexes relatively shallow.
**Habitat** shore down to 50m among algae and hydroids.
**Distribution** Mediterranean, Atlantic north to Norway, English Channel and North Sea.

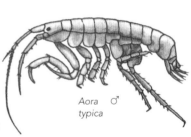
*Aora ♂ typica*

*Gammaropsis maculata*

### Gammaropsis maculata (Johnston)
**Length** up to 10mm.
**Head** bears 2 pairs of roughly equal antennae.
**Body** 1st and 2nd thoracic limbs have hairy, flattened and expanded 4th and 5th joints supporting claw-like 6th terminal joints. In the female the 5th joint of the 2nd limb has 2 teeth where it forms a pincer with the 6th joint; the male has 3 teeth.
**Habitat** from shore down to 250m.
**Distribution** Atlantic north to south-west Britain and Ireland.

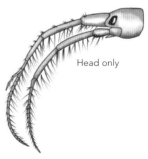
Head only

217

### Corophium volutator (Pallas)
**Length** up to 8mm.
**Head** upper antennae lacking a side-branch (which if present would be visible under water in a dish) and being about half the length of the lower antennae which are almost as long as the body, heavy and conspicuous.

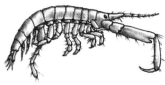

**Body** not compressed laterally like the preceding amphipods in this book; the last pair of apparently functional walking legs is much longer than the preceding pairs.
**Habitat** in u-shaped burrows in mud on the middle shore, estuaries and salt marshes.

*Corophium volutator*

**Distribution** Atlantic, English Channel, North Sea and Baltic.

### Jassa falcata (Montagu)
**Length** up to 8mm.

*Jassa falcata*

**Head** upper antennae have a very small side branch (visible when the animal is underwater in a dish) there are about three quarters the length of the lower antennae which appear noticeably thicker.
**Body** 1st thoracic leg terminates in a joint which is bent back as a pincer; 2nd thoracic leg does so too, but the pincer is larger and varies in shape from male to female (see inset illustration, right).
**Habitat** middle and lower shores; often with seaweeds on floating objects.
**Distribution** Mediterranean, Atlantic, English Channel and North Sea.

2nd thoracic
leg of *Jassa falcata*

### Chelura terebrans Philippi
**Length** up to 6mm.

**Head** upper antennae with a minute branch half way along (probably visible underwater in a dish with a hand lens); lower antennae twice as long with some what flattened, hairy joints.
**Body** not flattened side-ways; peculiar tail segment which is somewhat like the rudder of a jet-plane ending in a sharp point; of the last 3 pairs of abdominal appendages the 2nd is very conspicuous and long, exceeding the length of the tail segment and terminating in longish, oval plates, which are larger in the male than the female, and the 3rd is flattened into shorter plates.

*Chelura terebrans*

**Habitat** in bore holes in wood.
**Distribution** Mediterranean, Atlantic, English Channel and North Sea.

### Hyperia galba (Montagu)

**Length** female up to 20mm; male up to 12mm.
**Head** upper and lower antennae very small; eyes large and domed.
**Body** somewhat curved.
**Habitat** inside the umbrella of jellyfish (e.g. *Rhizostoma*, see page 87) in Summer, and living free on the seabed in Winter.
**Distribution** Mediterranean, Atlantic, English Channel, North Sea and west Baltic.

*Hyperia galba*

### Phtisica marina Slabber

**Length** up to 15mm.
**Head** upper antennae about twice as long as lower antennae.
**Body** long and thin; abdomen reduced and vestige of tail present.
**Habitat** associated with algae and other organisms in shallow water.
**Distribution** Mediterranean, Atlantic north to Scotland, English Channel and North Sea.

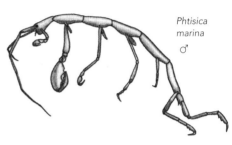

*Phtisica marina* ♂

### Caprella linearis (Linnaeus)
**Ghost Shrimp**

**Length** female up to 14mm; male up to 20mm.
**Head** upper antennae not branched and about twice as long as lower antennae, which have hairs all along them.
**Body** thin and long: abdomen reduced; vestige of tail present.
**Habitat** among hydroids, etc., on lower shore downward; sometimes swimming.
**Distribution** Atlantic, English Channel, North Sea and west Baltic.

*Caprella linearis* ♂

### Caprella acanthifera Leach

**Length** up to 9mm.
**Head** upper antennae considerably longer than lower; both antennae of male about twice as long as respective parts of female.
**Body** long and slender; 1st thoracic appendage pincered, long in male, short in female; female with broad pouch; thoracic segments bear dorsal spines (head not spiny); abdomen vestigial.
**Habitat** on hydroids, bryozoa etc.
**Distribution** Mediterranean, Atlantic north to Scotland, English Channel and North Sea.

*Caprella acanthifera* ♀

## ORDER DECAPODA

These malacostracans have the head and thorax fused and shielded by a carapace which bears a rostrum between the eyes, and the abdomen is clearly defined. There are eight pairs of thoracic appendages; the first and third pairs are developed as mouth parts, and the fourth to eighth pairs are used for walking and may terminate in a pincer or claw-like joint. Five pairs of abdominal appendages are used for swimming (and for brooding the eggs in the female), the last pair forming the tail fan on the terminal segment. The general malacostracan characteristics are shown on pages 206–207.

## INFRAORDER CARIDEA

### *PRAWNS AND SHRIMPS*

These decapods, which can swim, have light exoskeletons, and their bodies are sometimes flattened laterally. One pair of antennae is distinctly larger than the other which is more obviously branched at its base. The rostrum may be prominent (as in most prawns where its form is of use in identification), or greatly reduced (as in the shrimps). Further information is provided by Smalden, G. 1979.

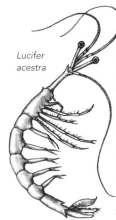

Lucifer
acestra

#### *Lucifer acestra* Dana

**Length** up to 10mm. Main antennae shorter than body; eye stalks conspicuously long; body flattened sideways; carapace small; thoracic legs delicate; swimmerets about half as long as thoracic appendages; tail with reduced fan.
**Colour** transparent.
**Habitat** pelagic.
**Distribution** Mediterranean.

#### *Palaemon elegans* Rathke
#### ( =*Leander elegans* or *L. squilla*)
**Prawn**

**Length** up to 50mm. Main antennae one and a half times as long as the body; rostrum may be slightly upcurved and terminates in 1 sharp tooth, between 7 and 10 teeth on the upper side occupy the full length of the rostrum, 3 teeth below situated together near the tip; front edge of carapace has 2 teeth on each side; the largest pincers occur on the 2nd pair of walking legs, and the 2nd, 3rd and 4th pairs are the longest.

Palaemon
elegans

**Habitat** in rock pools on the lower shore, often among seaweeds, also in shallow water.
**Distribution** Mediterranean, Atlantic, English Channel, North Sea and west Baltic.

### *Palaemon serratus* (Pennant) *( =Leander serratus)*
**Common Prawn**
(Not fully illustrated) Similar to *P. elegans* (opposite).

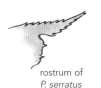

**Length** up to 65mm, sometimes greater, Rostrum curves upwards and terminates in 2 small equal teeth; there are 6–8 teeth on the upper edge, set back from the tip, and 4–5 below.
**Habitat** as for *P. elegans*.
**Distribution** Mediterranean, Atlantic and English Channel.

rostrum of
*P. serratus*

### *Palaemon adspersus* Rathke *( =Leander adspersus)*
**Prawn**
(Not fully illustrated) Similar to *P. elegans* (opposite).

**Length** up to 50mm. Rostrum slightly curved and terminating in 2 teeth, upper side has 5–7 teeth placed equally along its length, lower side has 3–4 grouped so that the 1st lies slightly in front of the 2nd tooth above.
**Habitat** as for *P. elegans*.
**Distribution** Mediterranean, Atlantic, North Sea and Baltic.

rostrum of
*P. adspersus*

### *Palaemon longirostris* Milne Edwards
**Length** up to 75mm. Main antennae about as long as the body, rostrum straight or only slightly curved upwards, 7 or 8 teeth on the upper side, 3, 4 or occasionally 5 on the lower side, 2 of the teeth on the upper side lie behind the eye socket. Largest pair of nippers on the 2nd walking leg. 2nd pair of legs are the longest.
**Habitat** in estuaries and also in brackish water.
**Distribution** Mediterranean and Atlantic north to south-west Britain.

*Palaemon longirostris*

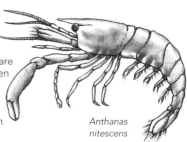

### *Anthanas nitescens* (Leach)
**Length** up to 20mm. Main antennae half as long as body; short untoothed rostrum ends in a sharp point; 1st pair of walking legs as long as the 2nd; pincers of the 1st pair are much larger, and the right pincer is often greater than the left.
**Habitat** in pools and among seaweeds on the lower shore and down to 70m.
**Distribution** Mediterranean, Atlantic, English Channel, North Sea and west Baltic.

*Anthanas nitescens*

### Alpheus glaber Milne-Edwards ( =Alpheus ruber)
**Snapping Prawn**

*Alpheus glaber*

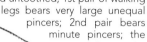

**Length** up to 35mm. Main antennae about as long as body; rostrum small and untoothed; 1st pair of walking legs bears very large unequal pincers; 2nd pair bears minute pincers; the remaining pairs terminate in a claw-like joint.
**Colour** pink-red.
**Habitat** usually from 30–100m, often on soft substrates and among growths.
**Distribution** Mediterranean, Atlantic and English Channel. N.B. by snapping its large pincers in the water this prawn produces audible vibrations which stun its prey.

### Hippolyte varians Leach
**Chameleon Prawn**

**Length** up to 25mm. Main antennae half as long as body; rostrum straight, as long as carapace, and terminating in a single point, 2 widely spaced teeth above and 2 below which are closer together; 3rd pair of walking legs is the longest.
**Colour** carapace variable depending on background – green, red or brown by day; transparent blue by night.
**Habitat** on the lower shore among rocks and seaweeds and down to 100m.
**Distribution** Mediterranean, Atlantic, English Channel and North Sea; not common from Scotland northwards.

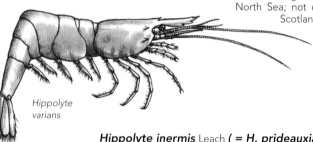

*Hippolyte varians*

### Hippolyte inermis Leach ( = H. prideauxiana)
Similar to *H. varians* (above).
**Length** up to 20mm. No teeth on the top of the rostrum which terminates in a single point, and has 2 teeth close together below.
**Habitat** among seaweeds and *Zostera* (see pages 64–65) on lower shore and in shallow water, sometimes in river mouths.
**Distribution** Mediterranean, Atlantic and west English Channel.

rostrum of *H. inermis*

### *Processa canaliculata* Leach

**Length** up to 74mm; carapace about one-third of body length; rostrum short, straight or slightly curved down, with 2 terminal teeth, the dorsal one shorter than the ventral; large eyes may appear higher than the rostrum; 2nd antennae longer than the body; unequal development of both 1st and 2nd pairs of walking legs.
**Colour** pinkish with red-orange patches; rostrum orange.
**Habitat** shallow water, often over sand down to deep water (200m).
**Distribution** Mediterranean and Atlantic north to Scotland, western English Channel and northern North Sea.

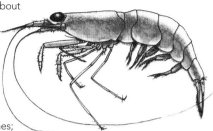

*Processa canaliculata*

### *Pandalina brevirostris* (Rathke)

**Length** about 20mm. Main antennae shorter than body; straight rostrum between a third and a half the length of the carapace, has about 5 teeth above and about 4 teeth below; carapace terminates in sharp point; 2nd pair of walking legs is the longest and bears pincers.
**Colour** shiny with reddish colours showing through.
**Habitat** in shallow water from 10–100m or more on clean, coarse gravel and sometimes among polyzoan colonies.
**Distribution** Mediterranean, Atlantic and English Channel.

*Pandalina brevirostris*

### *Pandalus montagui* Leach
**Aesop Prawn**

**Length** up to 40mm, sometimes more. Main antennae half as long again as body; rostrum as long as the carapace, curved and terminating in 2 teeth, has 10–12 teeth on the upper side, and 5–6 teeth on the lower side; 2nd pair of walking legs bears pincers, 2nd, 3rd and 4th pairs are the longest.
**Colour** carapace red-grey with darker spots and stripes; main antennae ringed with dark and light colours.
**Habitat** in rock pools on lower shore and in shallow water.
**Distribution** Atlantic, English Channel and North Sea.

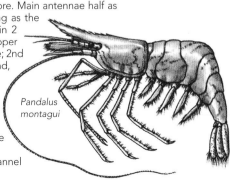

*Pandalus montagui*

### *Crangon crangon* Fabricius *( =C. vulgaris)*
**Common Shrimp**

**Length** up to 50mm, sometimes greater. Main antennae almost as long as the body; rostrum reduced to a small tooth; carapace has 1 central spine and 1 on either side; 1st pair of thoracic legs bear the biggest pincers, 2nd pair carry minute pincers, while 3rd and 4th pairs are the longest.

**Habitat** on the lower shore and in shallow water as well as in river estuaries.

**Distribution** Mediterranean, Atlantic, English Channel, North Sea and Baltic. N.B. The closely related shrimp *C. allmanni* (not illustrated) occurs off shore in all British waters. It is smaller and more brown in colour and like *C. crangon* is of commerical importance.

*Crangon crangon*

### *Crangon fasciatus* (Risso) *( =Philoceras fasciatus* or *Pontophilus fasciatus)*

**Length** up to 20mm. Main antennae shorter than the body; rostrum greatly reduced, as in *Crangon crangon* (above); carapace carries a noticeable spine on the upper surface; 1st pair of walking legs bears quite large but strangely formed pincers, 2nd pair with pincers, but very short, others end in a claw, 4th and 5th being the longest; abdominal appendages are relatively short.

**Habitat** on soft substrates sometimes with seaweeds from 4–40m.

**Distribution** Mediterranean, Atlantic and English Channel.

*Crangon fasciatus*

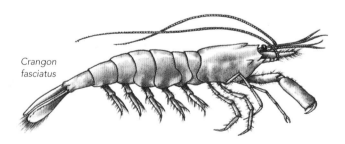

## Infraorder Astacidea
## Lobsters and their allies

Walking decapods which live between the tidemarks or on the seabed. They are strong and powerfully built often with heavy skeletons. The carapace has distinct transverse and oblique grooves running over it and is fused to the last thoracic segment. First three pairs of walking legs carry nippers (the first pair usually very large).

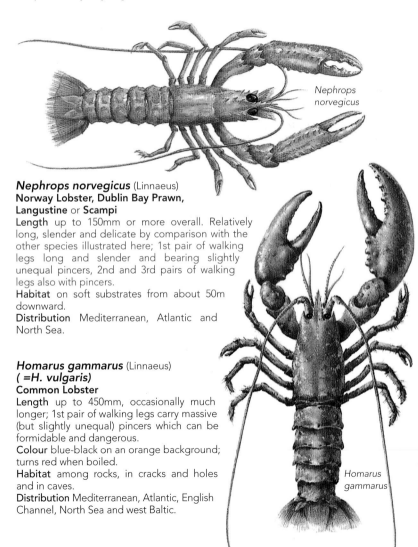

*Nephrops norvegicus*

### *Nephrops norvegicus* (Linnaeus)
**Norway Lobster, Dublin Bay Prawn,**
**Langustine** or **Scampi**
**Length** up to 150mm or more overall. Relatively long, slender and delicate by comparison with the other species illustrated here; 1st pair of walking legs long and slender and bearing slightly unequal pincers, 2nd and 3rd pairs of walking legs also with pincers.
**Habitat** on soft substrates from about 50m downward.
**Distribution** Mediterranean, Atlantic and North Sea.

### *Homarus gammarus* (Linnaeus)
**( =*H. vulgaris*)**
**Common Lobster**
**Length** up to 450mm, occasionally much longer; 1st pair of walking legs carry massive (but slightly unequal) pincers which can be formidable and dangerous.
**Colour** blue-black on an orange background; turns red when boiled.
**Habitat** among rocks, in cracks and holes and in caves.
**Distribution** Mediterranean, Atlantic, English Channel, North Sea and west Baltic.

*Homarus gammarus*

## Infraorder Palinura marine Crayfish or Crawfish and their allies

Similar to Astacidea but carapace very spiny, grooves less conspicuous, antennae very long and nippers much reduced or absent on first three pairs of walking legs. Small nippers sometimes on last (5th) pair.

### *Palinurus elephas* (Fabricius) *( =P. vulgaris)*
**Common Crawfish, Marine Crayfish, Spiny Lobster** or **Langouste**

**Length** 300–500mm overall. Easily recognized by the lack of pincers on any walking legs (except the 5th pair in the female which does have them); abdominal segments are sharply spined and can cause bad wounds if the animal is mishandled.

**Colour** red, brown or shaded with plum colour.

**Habitat** among rocks and in crevices, occasionally on stony substrates; from shallow water to 70m and deeper.

**Distribution** Mediterranean, Atlantic and English Channel. N.B. The small European freshwater crayfish *Astacus* should in no way be confused with this species!

*Palinurus elephas*

### *Scyllarus arctus* (Linnaeus)
**Shovel-nosed 'Lobster'**

**Length** up to 150mm overall. Second antennae relatively more broad and produced into about 5 easily recognized processes along the front edge; walking legs lack pincers (except for the 5th pair in the female which does have them); abdominal segments are more rounded at their edges than in preceding species.

**Habitat** among rocks with mud and on stony bottoms from 3m downward.

**Distribution** Mediterranean and Atlantic north to English Channel approaches.

*Scyllarus arctus*

## INFRAORDER ANOMURA
## Squat lobsters, hermit crabs, etc.

Walking decapods which live on the shore or seabed. The abdomen is not greatly reduced, but is often twisted, as in the hermit crabs, or folded under the thorax, as in the squat lobsters. The last thoracic segment is free from the carapace.

### Galathea intermedia Liljeborg
**Length** up to 10mm. Rostrum tip (see diagram, below) slightly blunt, tip itself longer than the 4 small spikes on the side of the rostrum; 1st walking leg bears pincers and is about twice as long as the body.

Rostrum of *G. intermedia*

*Galathea intermedia*

**Colour** bright red with blue spots.
**Habitat** among stones and rocks on lower shore and down to 80m.
**Distribution** Mediterranean, Atlantic, English Channel and North Sea.

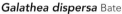

### Galathea dispersa Bate
**Length** up to 45mm. Rostrum covered with hairs and with 4 points on the sides; 1st walking leg bears hairy pincers and is as long again as body.
**Colour** dull orange or red; may be patterned.
**Habitat** among stones and rocks on lower shore and down to about 60m.
**Distribution** Atlantic, English Channel and North Sea.

*Galathea dispersa*

### Galathea squamifera Leach

*Galathea squamifera*

**Length** up to 45mm. Rostrum has 4 pairs of side-points and the hindmost pair is the smallest, the remaining side-points being all about the same size (see diagram, right); 1st walking leg one and a half times as long as body and bears pincers; pincer joints themselves have scales and spines on their outer edges, but the joints nearer the body have scales and spines on their inner or opposite edges.

Rostrum of *G. squamifera*

**Colour** generally green-brown, sometimes with a little red.
**Habitat** under stones and rocks from lower shore down to 80m.
**Distribution** Mediterranean, Atlantic, English Channel and North Sea.

### *Galathea strigosa* (Linnaeus)
**Length** up to 120mm, but often smaller. Rostrum pointed with 3 pairs of spines; 1st walking leg bears pincers about one and a half times as long as body; pincers and the 1st 3 pairs of legs are spiny.

> **Colour** red with blue lines across it.
> > **Habitat** under stones and rocks from lower shore down to 35m.
> > **Distribution** Mediterranean, Atlantic, English Channel and North Sea. N.B. aggressive when handled.

*Galathea strigosa*

### *Munida rugosa* (Fabricius)
**Length** up to 60mm. Antennae not quite as long as pincers and the appendage which bears them (in most squat lobsters antennae are as long or longer than pincer-bearing legs); 1st pair of walking legs may be twice as long as body and bears delicate, long, thin pincers.

> **Habitat** normally in deep water on sandy or other soft substrates from 50–150m.
> > **Distribution** Mediterranean, Atlantic and North Sea.

**Note** several groups of anomurans occur apart from the squat lobsters. Some of them resemble other divisions, such as the lobsters and true crabs, but their affinities with the squat lobsters and hermit crabs are indicated by their long and conspicuous antennae and by the miniature fifth pair of walking legs which is a key character of the porcelain 'crabs'. Crothers, J. and Crothers, M. 1983 and Ingle, R. W. and Christiansen, M. E. 2004 will also help with these groups.

*Munida rugosa*

*Porcellana platycheles*

### *Porcellana platycheles* (Pennant)
**Broad-clawed Porcelain 'Crab'**
> **Length** up to 12mm. Carapace round and squat; 1st pair of walking legs about 20mm long with thick and conspicuous pincers which are hairy on their outer edges; other legs hairy; abdomen folded tightly and out of sight under the carapace.
> > **Colour** yellow-brown with dirty grey or reddish appearance.
> **Habitat** under stones among mud and gravel on mid-

dle shore down to shallow water.
**Distribution** Mediterranean, Atlantic, English Channel and
North Sea. N.B. not a true crab.

### *Pisidia longicornis* (Linnaeus)
**Long-clawed Porcelain 'Crab'**
Similar to *P. platycheles* (opposite) in overall appearance.
**Length** up to 60mm. Antennae longer than carapace;
carpace round; pincers on the 1st walking legs long
and relatively slender; animal is not hairy, but relatively
'clean' in appearance.
**Colour** brown-red.
**Habitat** under stones and in *Laminaria* hold-fasts
(see pages 32–33) on lower shore and in shallow water.
**Distribution** Mediterranean, Atlantic, English Channel and
North Sea.

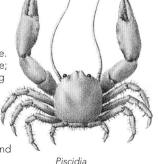

*Piscidia
longicornis*

### *Jaxea nocturna* (Chiereghin) Nardo
Superficially prawn-like.
**Length** about 50mm. Eyes not apparent; pointed rostrum; 1st
pair of walking legs bears strangely shaped, long, hairy pincers;
remaining legs are slender and frail-looking, being slightly
hairy; 2nd pair has small incomplete pincers; abdomen may
be folded under thorax and terminates in a tail fan.
**Colour** white, pink or brown.
**Habitat** in mud from about 15m downward.
**Distribution** common in Adriatic, rare in Mediterranean,
Atlantic and English Channel.

*Jaxea
nocturna*

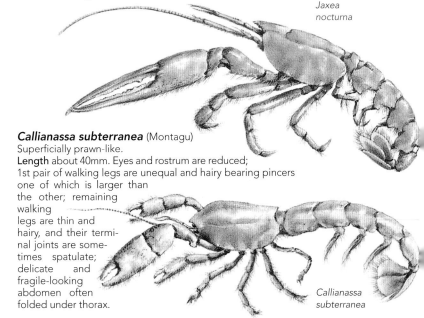

### *Callianassa subterranea* (Montagu)
Superficially prawn-like.
**Length** about 40mm. Eyes and rostrum are reduced;
1st pair of walking legs are unequal and hairy bearing pincers
one of which is larger than
the other; remaining
walking
legs are thin and
hairy, and their termi-
nal joints are some-
times spatulate;
delicate and
fragile-looking
abdomen often
folded under thorax.

*Callianassa
subterranea*

*Upogebia
deltaura*

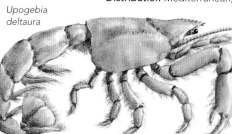

Colour white to pale red or bluish.
Habitat burrows in sand or mud from about 2m downward.
Distribution Mediterranean, Atlantic and English Channel.

### *Upogebia deltaura* (Leach)
Superficially prawn-like.
Length up to 100mm.
Head small; hairy rostrum; eyes reduced; 1st walking leg bears strange pincer with moving part longer than fixed part; other walking legs lack pincers, all are hairy; slender, delicate-looking abdomen may be folded under the thorax.
Colour white-grey-yellow-green.
Habitat burrowing in clay and muddy sand from lower shore to deep water.
Distribution Mediterranean, Atlantic, English Channel and North Sea.

### *Axius stirhynchus* Leach
Superficially prawn-like.
Length overall up to 72mm; carapace bears a short flat triangular rostrum and is noticeably compressed side-ways, being widest in the middle and tapering slightly towards the anterior and posterior; eyes small; long outer antennae more than twice the length of the inner ones; 1st pair of walking legs bears massive unequal pincers; 2nd pair has small flattened pincers; abdomen slender with wide tail fan.
Colour pale red-brown.
Habitat in shallow water or even on the shore, burrowing in mud and soft deposits.
Distribution Mediterranean and Atlantic north to south-west Britain.

*Axius
stirhynchus*

### Hermit crabs
The Anomura includes a number of hermit crabs which are familiar seashore animals. They are easily recognized because of their association with empty gastropod shells which protect their delicate, soft-skinned abdomens. Sponges, hydroids and sea-anemones often grow on the outside of shells inhabited by hermit crabs.

### *Diogenes pugilator* (Roux)
Length up to 25mm; carapace about 10mm (animal may be somewhat smaller). Hairy antennae; 1st pair of walking legs terminates in pincers, the left being the larger; 2nd and 3rd walking legs bear claws; 4th and 5th are greatly reduced.

*Diogenes
pugilator*

Colour eyes black; pincers with chalk-white tips.
Habitat in gastropod shells usually in shallow
water.
Distribution Mediterranean, and Atlantic
north to English Channel approaches.

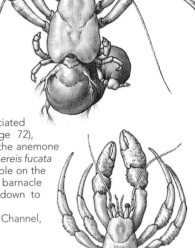

*Pagurus
bernhardus*

### *Pagurus bernhardus* (Linnaeus)
**Common Hermit Crab**
Length up to 100mm; carapace up to
40mm. First pair of walking legs bears
large, unequal, coarsely granulated pincers, the right being the larger; 2nd and
3rd walking legs terminate in claws
which are spiny; 4th and 5th walking
legs greatly reduced.
Colour carapace grey-red; pincers
red-brown.
Habitat in gastropod shells, sometimes associated
with sponges such as *Suberites* (see page 72),
hydroids such as *Hydractinia* (see page 80), the anemone
*Calliactis* (see page 97), and the polychaete *Nereis fucata*
(see page 121); a yellowish lump may be visible on the
side of the abdomen – this is the parasitic barnacle
*Peltogaster paguri;* from the lower shore down to
deeper water.
Distribution Mediterranean, Atlantic, English Channel,
North Sea and west Baltic.

### *Pagurus prideauxi* (Leach)
Length up to 60mm; carapace nearly 20mm. First pair of
walking legs bears slightly bristled pincers with fine granules,
the right being slightly larger; 2nd and 3rd pairs of walking
legs bear claws which are grooved but not spiny; 4th and 5th
pairs are greatly reduced.
Colour carapace brown-red.
Habitat in small shells often associated with the cloak
anemone *Adamsia carcinopados* (see page 97); found on
mud or sand from 10m downward.
Distribution Mediterranean, Atlantic, English Channel and
North Sea.

*Pagurus
prideauxi*

### *Anapagurus laevis* (Thompson)
Length up to 20mm; carapace almost 10mm. First pair of
walking legs bears slightly hairy pincers, the right being the
larger; 2nd and 3rd walking legs are slender and bear claws;
4th and 5th pairs are greatly reduced.
Colour carapace yellow-white; pincers banded with orange.
Habitat in small gastropod shells from 10m downward.
Distribution Mediterranean, Atlantic, English Channel and
North Sea.

*Anapagurus
laevis*

231

*Pagurus cuanensis*

Detail of pincer

### *Pagurus cuanensis* Thompson ( =*Eupagurus cuanensis)*

**Length** up to 25mm, carapace up to 6mm long; upper surface of right pincer is hairy and bears granules.
**Colour** yellow-brown.
**Habitat** from 10–100m.
**Distribution** Mediterranean and Atlantic north to Norway, English Channel and North Sea.

### *Clibanarius erythropus* Latreille

**Length** up to 20mm, carapace about 5mm, often smaller; antennae not hairy; 1st pair of walking legs bears pincers of roughly equal size.
**Colour** tips of walking legs and pincers black, remainder of these limbs spotted or striped pale blue or red; eye-stalks red; body red-brown-green.
**Habitat** in shallow water under stones and on gravel, often very common.
**Distribution** Atlantic north to south-west Britain.

*Clibanarius erythropus*

## INFRAORDER BRACHYURA
## True crabs

Walking decapods which live on the shore or seabed. True crabs are usually strong and powerfully built with a heavy skeleton. The carapace is flattened and rounded, forming the typical crab-shaped body, and the abdomen is much reduced and folded forwards under the carapace. The antennae are usually short. The first pair of walking legs bears conspicuous and often powerful pincers, while the other four pairs of walking legs are variously developed, ending normally in a claw-like joint. Further information on European and British crabs is given by Crothers, J. and Crothers M. 1983, Ingle, R. W. 1980 and 1996.

*Dromia personata*

### *Dromia personata* (Linnaeus)
### Sponge Crab

**Length** up to 80mm. Carapace slightly broader than it is long, and domed; whole body, including appendages, covered for the greater part by hairs which give the crab a furry appearance; 4th and 5th pairs of legs displaced, and the 5th pair appears to be carried on the crab's back; 5th pair terminates in small pincers.
**Colour** hairs dark brown; tips of pincers bright pink.
**Habitat** on sandy and rocky shores from low water down to 30m.

**Distribution** Mediterranean, Atlantic and English Channel.
N.B. often found carrying a piece of sponge on its back.

### Ebalia nux Norman

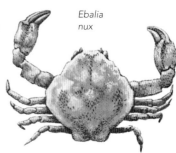

*Ebalia
nux*

**Length** up to 8mm. Carapace rounded, width to
length ratio abou 1:1; small central notch in
leading edge and 2 small teeth either side
between notch and eye; upper surface smooth
and lacking spines or knobs; 1st pair of walking
legs long and bearing slender pincers.
**Colour** pinkish, reddish or brown.
**Habitat** on muddy bottoms below 80m.
**Distribution** Mediterranean, Atlantic north to
Scotland.

### Ebalia cranchi Leach

**Length** up to 7mm. Carapace granulated, with a rhomboidal
outline when viewed from above; a notch is present in the
posterior and anterior corners and bears 5 conspicuous
knobs, 1 posteriorly, 2 laterally ( 1 left and 1 right) and 2
almost at the centre.
**Colour** carapace yellow-red.
**Habitat** on sand and gravel from 20–130m; often
uncommon.
**Distribution** Mediterranean, Atlantic, English Channel
and North Sea. N.B. at least 4 other species of *Ebalia* have
been recognized from European waters all of which have
somewhat rhomboidal carapaces, but differ in the presence or
absence of granules and knobs, as well as in the shape and
proportions of the 1st walking legs.

*Ebalia
cranchi*

### Calappa granulata (Linnaeus)

**Length** up to 110mm. Carapace almost oval in outline, being
broader than it is long; front edge strongly convex and finely
toothed, rear edge weakly convex with fewer, larger
teeth; 1st pair of walking legs bear
strongly built pincers which have
a crest on their upper side
reminiscent of a cock's comb.
**Colour** carapace light grey to
yellow, spotted with red.
**Habitat** on muddy substrates
from 30–100m.
**Distribution** Mediterranean and
Atlantic north to Portugal.

*Calappa
granulata*

233

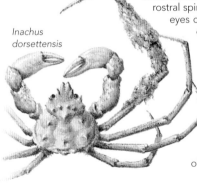

Macropodia
tenuirostris

### *Macropodia tenuirostris* (Leach)
**Spider Crab**
**Length** up to 19mm. Carapace triangular, not as wide as it is long; bears 7 thorny spikes and is often encrusted with sponges and seaweeds; rostrum prominent and long being composed of 2 parallel spines joined all along; eyes borne on either side of rostrum, but not retractable into their sockets; antennae have bristles at their joints; 1st pair of walking legs bears pincers, other walking legs long and slender; 2nd and 3rd pairs are the longest and terminate in a more-or-less straight claw, 4th and 5th pairs shorter terminating in a more-or-less curved claw with teeth on inner surface; legs slightly hairy.
**Colour** carapace yellow-red.
**Habitat** in shallow water camouflaged among seaweeds, etc., and in deeper water.
**Distribution** Mediterranean and Atlantic, English Channel, North Sea and west Baltic.

### *Macropodia rostrata* (Linnaeus)
**Spider Crab**
(Not illustrated) Similar to *M. tenuirostris* (above).
**Length** up to 22mm. Carapace triangular-shaped, with about 8 conspicuous spines; rostrum much shorter; antennae lacking bristles at their joints.
**Colour** carapace yellow, grey, green or brownish.
**Habitat** on hard bottoms from 4–90m.
**Distribution** as for *M. tenuirostris.*

### *Inachus dorsettensis* (Pennant)
**Spider Crab**
**Length** about 25mm. Carapace triangular, about as wide as it is long, bears 4 small tubercles arranged in a row across the front end with one large tubercle behind them; 2 large tubercles at the back end with a smaller tubercle between them; 2 rostral spines between the eyes not joined together; eyes can be retracted into their sockets; 1st pair of walking legs bears relatively stout pincers. 2nd pair is about three times the shell length, and the remaining pairs become progressively smaller towards the rear.
**Colour** carapace yellow-brown.
**Habitat** on stony substrates and among seaweeds on lower shore, in shallow water and down to about 100m.
**Distribution** Mediterranean, Atlantic, English Channel and North Sea. N.B. often covered with sponges and pieces of seaweed.

Inachus
dorsettensis

### *Pisa armata* (Latreille)
Length up to 40mm. Carapace more or less triangular; rough-topped, slightly domed, and with slightly concave sides; not quite as wide at the rear as it is long; 2 rostral spines (parallel in the male; but diverging in the female), with another smaller spine in front of each eye; usually 3 posterior tubercles, one pointing backward, and one to either side; 1st pair ot walking legs bears strong, but not large pincers; remaining pairs become slightly shorter towards the posterior.
**Colour** carapace brown.
**Habitat** usually below 20m.
**Distribution** Mediterranean, Atlantic and English Channel. N.B. may be encrusted with seaweeds, sponges or anemones.

*Pisa armata*

### *Maja squinado* (Herbst)
**Spiny Spider Crab**
Length up to 180mm. Carapace not as wide posteriorly as it is long, somewhat triangular, but with convex, rounded outline marked by large and small spines; covered with spines and bristles; 2 spines between eyes; 1st pair of walking legs fairly long with small, equal pincers, remaining legs hairy and long except the 5th pair.
**Colour** carapace red, pink or white, sometimes spotted.
**Habitat** on sand and among rocks from the lower shore down to about 50m.
**Distribution** Mediterranean and Atlantic north to the English Channel. N.B. may be encrusted with seaweeds or sponges.

*Maja squinado*

### *Hyas araneus* (Linnaeus)
**Spider Crab**
Length up to 110mm. Carapace somewhat pear-shaped, slightly domed and covered with small tubercles and bristles; 2 rostral spines lie close together to make a triangular rostrum, on either side of which is a smaller spine; 1st pair of walking legs bears pincers, remaining legs of similar length.
**Colour** carapace dull red to dull brown.
**Habitat** among rocks and on sand, usually in shallow water and sometimes among seaweeds.
**Distribution** Atlantic from the English Channel northwards, North Sea and Baltic. N.B. may be encrusted with seaweeds, sponges and hydroids.

*Hyas araneus*

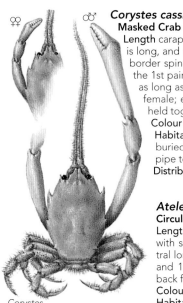

### *Corystes cassivelaunus* (Pennant)
**Masked Crab**

**Length** carapace up to 40mm. Carapace not as broad as it is long, and relatively smooth; 1 small and 1 conspicuous border spine occur between the eyes and the origin of the 1st pair of walking legs, which themselves are twice as long as the carapace in the male, but shorter in the female; exceptionally long, hairy antennae which are held together throughout their length.

**Colour** carapace drab brown-yellow.

**Habitat** lower shore and in shallow water, usually buried in sand, using the joined antennae like a pipe to allow water to reach the gills.

**Distribution** Atlantic, English Channel and North Sea.

*Corystes
cassivelaunus*

### *Atelecyclus rotundatus* (Olivi)
**Circular Crab**

**Length** up to about 30mm, carapace almost round with some fine surface granulations, 3 teeth, central longest at front between antennae, between 9 and 11 teeth on either side of carapace running back from eye socket.

**Colour** red-brown, legs paler, nippers black.

**Habitat** on sand and gravel from very shallow water down to 300m.

**Distribution** Atlantic north to Norway, English Channel and North Sea.

*Atelecyclus
rotundatus*

### *Cancer pagurus* Linnaeus
**Edible Crab**

**Length** up to 140mm, frequently smaller and very occasionally larger. Carapace is lightly granulated and slightly domed, being half as wide again as it is long, oval in outline with about 9 rounded lobes on the edge of each side giving the animal an appearance reminiscent of a piecrust; 1st pair of walking legs bears massive pincers; 4 remaining pairs of walking legs are hairy, the 5th pair being the smallest.

**Colour** carapace pink to brownish; pincers tipped with black.

**Habitat** from lower shore down to about 100m, usually among rocks; really large specimens are normally found in deep water.

**Distribution** Mediterranean, Atlantic, English Channel and North Sea.

*Cancer
pagurus*

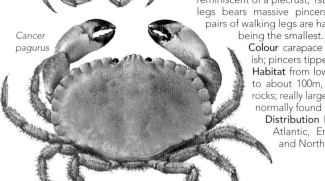

### *Pirimela denticulata* (Montagu)

**Length** about 25mm. Carapace bears 3 small projections between the eyes, the outer 2 being triangular and flattened; the middle longer and round; 7 more teeth developed on either side of the body; 1st pair of walking legs bears small pincers; all the other pairs have terminal joints which are pointed and not paddle-like.
**Colour** carapace variable; green, brown, purple-red; may be mottled.
**Habitat** rocky places or with gravel and seaweeds from lower shore down to about 60m.
**Distribution** Mediterranean, Atlantic, English Channel and North Sea.

*Pirimela denticulata*

### *Carcinus maenas* (Linnaeus)
**Common Shore Crab**

**Length** up to 40mm, occasionally larger. Carapace up to half as wide again as it is long; 3 blunt teeth between the eyes; 5 sharp, well-developed teeth on either side of the body; 1st pair of walking legs bears moderately sized, powerful pincers, 2nd and 3rd pairs are the longest, 5th is the shortest and the terminal joints are flattened, but pointed.
**Colour** carapace brown, olive or dark green above, green-yellow below.
**Habitat** on sandy and rocky shores and in shallow water often where salinity is variable and in polluted places.
**Distribution** Mediterranean, Atlantic, English Channel, North Sea and Baltic.

*Carcinus maenas*

### *Liocarcinus depurator* (Linnaeus)
**Swimming Crab**

**Length** up to 40mm. Carapace slightly wider than it is long; carapace sometimes rough and scaly with some hairs; 3 sharp teeth between the eyes which are about the same size as the 5 on either side of the shell (sometimes there is an extra, smaller point at the inner corner of the eye socket); 1st pair of walking legs bears pincers, 5th pair has the last joint flattened and rounded into a paddle for swimming (not pointed as in *Carcinus*).
**Colour** carapace reddish brown.
**Habitat** often on sandy substrates from 5–20m.
**Distribution** Mediterranean, Atlantic, English Channel and North Sea. N.B. several other species occur.

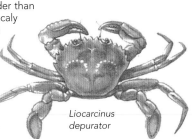

*Liocarcinus depurator*

*Necora puber*

### *Necora puber* (Linnaeus)
### Velvet Swimming Crab

**Length** about 80mm or less. Carapace bears 8–10 small teeth between eyes of which the middle 2 are longest; 5 large, pointed teeth on either side of the edge of the shell towards the front; 1st pair of walking legs bears strong pincers; last joint of back legs is flat and rounded to act as a swimming paddle (not pointed as if *Carcinus*, page 237).
**Colour** carapace red-brown; covered with fine hair giving a muddy-brown appearance; eyes red.
**Habitat** among stones and rocks from the lower shore down to about 10m.
**Distribution** Atlantic, English Channel and North Sea N.B. may defend itself strongly when disturbed.

### *Pilumnus hirtellus* (Linnaeus)
### Hairy Crab

*Pilumnus hirtellus*

**Length** up to 20mm. Carapace anterior edge formed into 2 shallow lobes between the eyes; 5 points on the edge outside the eyes (including the eye-socket margin); carapace is wider than it is long, rounded in front, but tapering posteriorly; large, strong, unequal pincers on the 1st walking legs; whole animal is hairy.
**Colour** carapace brownish red; pincers brown.
**Habitat** on rocks and stones, in cracks and holes, sometimes among sponges; lower shore and in shallow water.
**Distribution** Mediterranean, Atlantic and English Channel.

### *Xantho incisus* Leach
### Montagu's Crab

**Length** about 20mm. Carapace broad in front and tapering sharply to the posterior; shallow furrow runs in from between the eyes, 2 indentations on each side of the carapace; 1st pair of walking legs carries relatively large, unequal pincers; other legs become smaller towards the rear.
**Colour** carapace variable; yellow, brown, green or red-brown.
**Habitat** among sand, gravel and loose stones on lower shore and in shallow water.
**Distribution** Mediterranean, Atlantic and English Channel.

*Xantho incisus*

### Pinnotheres pisum Linnaeus
**Pea Crab**

*Pinnotheres pisum*

**Length** of female carapace up to 13mm, width up to 14mm; length of male carapace up to 6mm, width up to 6mm. Female carapace soft, rounded and smooth, 1st pair of walking legs bears quite delicate pincers; last joint of 5th walking legs relatively short and hook-like. Male carapace hard, leading edge extended slightly between eyes, pincers of 1st pair of walking legs stronger; all legs stronger and more hairy than female; hook-like last joint to 5th pair.
**Colour** female transparent, brown-yellow above bearing a yellow spot at front and yellowish patches at side; male yellow-grey.
**Habitat** inside bivalve shells e.g. *Mytilus* and *Spisula*.
**Distribution** Mediterranean, Atlantic, English Channel and North Sea. N.B. similar species include *P. pinnotheres* (Linnaeus) (not illustrated) with terminal joint of 5th walking leg long, straight and tapering, about same length as terminal joint.

### Goneplax rhomboides (Linnaeus)
**Angular Crab**

**Length** up to 27mm, often less. Carapace width to length ratio 7:4; carapace oblong, leading edge reasonably straight and drawn out into conspicuous teeth, with a second pair a little behind on either side; eyes carried on long eyestalks often folded back against front of carapace; in the male the 1st pair of walking legs is very long, the pincer-bearing joint is about equal to carapace length; other legs about equal.
**Colour** yellow-red.
**Habitat** on soft substrates down to 100m.
**Distribution** Mediterranean, Atlantic, English Channel and North Sea.

*Goneplax rhomboides*

### Eriocheir sinensis Milne-Edwards
**Chinese Mitten Crab**

**Length** up to 70mm. Carapace more or less square and wide-fronted, with 4 teeth and a shallow groove between the eyes; 1st pair of walking legs bears quite powerful pincers whose parts are invested with 'fur' giving the appearance of mittens; other legs long and hairy.
**Colour** carapace olive-green; sometimes marked with blotches.
**Habitat** in fresh water and estuaries.
**Distribution** Atlantic north from the English Channel, North Sea and Baltic.

*Eriocheir sinensis*

# PHYLUM MYRIAPODA

## CLASS CHILOPODA Centipedes

Long, flattened arthropods with many similar body segments, and not more than one pair of appendages per segment. The head has antennae and simple eyes, and the first body segment bears poisonous claws.

**Strigamia maritima** (Leach)
**Length** up to 40mm.
**Habitat** under stones and in rock crevices on the upper shore.
**Distribution** Atlantic, English Channel, North Sea and west Baltic.

*Strigamia maritima*

# PHYLUM UNIRAMIA Insects

Arthropods in which the adult body is divided into three distinct regions: a head with a single pair of antennae; a thorax with three pairs of legs and often one or two pairs of wings; and an abdomen without walking appendages.

*Petrobius maritumus*

*Lipura maritima*

**Petrobius maritimus** (Leach)
**Bristle-tail**
**Length** up to 12.5mm.
**Head** conspicuous antennae which are about the length of the body; it also has a pair of palps.
**Body** thorax bears 3 pairs of legs but no wings; abdomen terminates in a long bristle which itself is nearly as long as the body.
**Habitat** upper shore and landward in rocky crevices and cracks.
**Distribution** Mediterranean, Atlantic, English Channel and North Sea.

**Lipura maritima** (Laboulbène)
**Length** about 3mm.
**Head** bears short antennae.
**Body** thorax and abdomen quite plump; thorax bears 3 pairs of short legs, but no wings; abdomen broad towards the tail, but tapering to a blunt point at the tip.
**Habitat** generally floating on the surface film of water in rock pools on the upper shore, or crawling on rocks and seaweeds.
**Distribution** Mediterranean, Atlantic, English Channel and North Sea.

## CLASS ARACHNIDA
### ORDER PSEUDOSCORPIONES *FALSE SCORPIONS*

*Neobisium maritimum*

The adult body is made up of two primary divisions: a cephalothorax (head and thorax) of six segments bearing conspicuous pincers; and a pair of appendages with fangs as well as 4 pairs of walking legs.

### *Neobisium maritimum* (Leach)
**Length** 2mm.
**Body** with the aid of a hand lens the characteristic pincers can easily be seen on the cephalothorax.
**Habitat** generally found in rock crevices and under other cover on upper shore.
**Distribution** Atlantic, English Channel and North Sea.

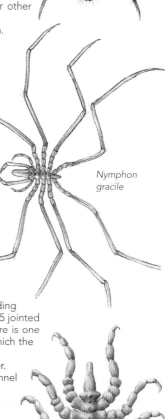

## CLASS PYCNOGONIDA
## (SOMETIMES REGARDED AS A SUBPHYLUM)
## Sea-spiders

Exclusively marine arthropods with a cephalothorax (head and thorax) drawn out anteriorly into a proboscis which opens by a terminal mouth. The abdomen is reduced to a single segment. Four pairs of relatively long legs are borne by the thorax, into which the ovaries and digestive system extend. King, P. E. 1974 gives a fuller account of the pycnogonids.

*Nymphon gracile*

### *Nymphon gracile* Leach
**Length** up to 10mm, sometimes longer; walking legs reach up to 25mm.
**Body** slender cephalothorax; proboscis equipped on either side with a pair of pincer-like feeding appendages (chelicerae), behind which are a pair of 5 jointed palps; in addition to the 4 pairs of walking legs there is one pair of 'ovigerous' legs (held below the body) on which the male carries the eggs after the female has laid them.
**Habitat** middle and lower shores and in shallow water.
**Distribution** Mediterranean, Atlantic, English Channel and North Sea.

### *Pycnogonum littorale* (Ström)
**Length** up to 20mm.
**Body** no palas or chelicerae, ovigerous legs only in males; cephalothorax bears prominent proboscis; relatively heavy body with thick legs.
**Habitat** lower shore under stones, etc.
**Distribution** Atlantic, English Channel and North Sea.

*Pycnogonum littorale*

# PHYLUM ECTOPROCTA ( =BRYOZOA OR POLYZOA)

Ectoprocts are minute, sessile, colonial animals. Individuals grow inside a secreted case called the *zooecium*. A true body cavity (coelom) is present, and the mouth is surrounded by a ring of hollow, ciliated tentacles (collectively called a lophophore), which can be retracted inside the zooecium. The anus lies outside the lophophore.

Ectoprocts are among the commonest animals inhabiting stony and rocky shores and seabeds, yet they are often neglected – probably on account of their small size and lack of commercial value. They are divided into three classes: Phylactolaemata, Stenolaemata and Gymnolaemata, of which the first class is exclusively freshwater and the second consists largely of fossil forms with but one living order. The body plan is quite characteristic. Individuals dwell in a colony which has been developed by asexual budding from one ancestral animal. Colonies such as *Membranipora* often encrust rocks, shells or seaweeds, but other ectoprocts, like *Flustra*, grow largely unsupported, being attached only at their bases. The zooecium is sometimes hardened by chalky secretions from the body wall; and its shape may be flat and box-like, or tubular. The mouth leads to a u-shaped gut lying inside the zooecium, where the reproductive organs are also housed. Circulatory and excretory structures appear to be lacking.

The general form of the ectoproct animal is shown below left.

One striking feature of this phylum is the evolution of polymorphic-individuals within a colony. Many members of the colony may be feeding individuals, but some are specialized to fulfil other roles. One type called *avicularia* resemble minute birds beaks. They appear to defend the other members from small organisms which would otherwise creep over them and possibly clog them up. *Vibracula* are another variety of zooid; these bear miniature paddles which probably aid the circulation of water around the colony and discourage the accumulation of silt and other particles. The principal type of feeding individuals filter small food particles from the surrounding sea water by means of the ciliated tentacles of the lophophore. The diagram on page 243 is a magnified view of a colony as seen under the microscope and shows the form of part of the colony as well as feeding zooids and avicularia. In the illustrations that follow the overall growth pattern and shape of the colony is shown as well as a magnified view of zooids enclosed in a ring to

▼ Diagrammatic arrangement of an individual or zooid

- Hollow tentacle of lophophore
- Position of mouth
- Anus
- Nerve ganglion
- Gut
- Ovary
- Body cavity
- Body wall
- Zooecium
- Testis
- Substrate

indicate the field of the microscope.

Although each colony develops by budding from an ancestral individual, sexual reproduction leads to the development and dispersal of free larvae. After a period in the plankton these larvae metamorphose into new ancestral individuals, assuming they can find a suitable place on which to settle, and then develop a new colony by asexual budding. In some species the developing embryos are brooded either inside the body or in special pouches called *ooecia*. Colonies usually contain both male and female zooids, but sometimes the zooids themselves are hermaphrodite. Ectoprocts do not generally flourish in brackish water, but a few species, for instance *Electra crustulenta*, can tolerate low salinities.

Some clue to the identification of an ectoproct colony will immediately be given by its overall appearance – e.g. erect or branching; flat or encrusting, etc. – but a good hand lens will be essential in order to show the fine details of the zooids themselves which are necessary to establish final, identification. The general shape of the zooecium, the position of the opening through which the lophophore is protracted, the presence of an operculum (effectively a lid) for occluding the opening, the presence of ooecia, aviculariae, etc., are all very important in helping to make an identification. Some species also carry spines and bristles which also assist with their identification. Unfortunately, many more species of ectoproct occur in the European area than can be descibed here. Comprehensive references for the identification of ectoprocts are: Hinks, T. 1880; Ryland, J. S. 1962; Ryland, J. S. 1974; Ryland, J. S. and Hayward. P. J. 1977; Hayward, P. J. 1985; Hayward, P. J. and Ryland, J. S. 1985, 1998 and 1999; Ryland, J. S. 1970 gives an excellent account of the biology of ectoprocts.

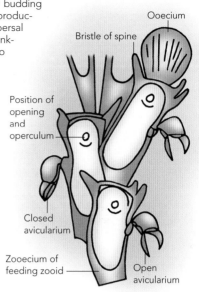

Ooecium

Bristle of spine

Position of opening and operculum

Closed avicularium

Zooecium of feeding zooid

Open avicularium

▲ Arrangement of a few individuals of *Bugula*

# Class Stenolaemata

Ectoprocts with tubular zooecia, and with the zooecial walls calcified. The lophophore is often circular. Terminal opening not closed by an operculum.

## Order Cyclostomata

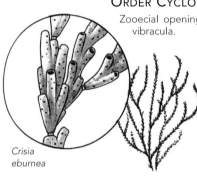

Zooecial opening rounded. No operculum, avicularia nor vibracula.

*Crisia*
*eburnea*

### Crisia eburnea (Linnaeus)

Branched, jointed colony up to 20mm high; attached at base; a long spine grows behind each zooid.

**Colour** grey-white.

**Habitat** growing on red seaweeds, shells and rocks from middle shore down to 50m.

**Distribution** Mediterranean, Atlantic, English Channel and North Sea.

# Class Gymnolaemata

Zooecia tubular and box-like; the zooecial walls may be calcified. The lophophore is circular.

## Order Ctenostomata

Zooecial walls not calcified; zooecial opening may be shut by a collar. No operculum, ooecia, avicularia nor vibracula are present.

*Bowerbankia*
*imbricata*

### Bowerbankia imbricata (Adams)

Tufted, branching colony up to 70mm high; attached at base; zooids at intervals along stems, often in groups.

**Colour** grey-buff; golden embryos developing in zooecia may be seen with a hand lens.

**Habitat** growing on seaweeds, e.g. *Ascophyllum* and *Fucus* (see pages 38–39) on middle and lower shore and in shallow water.

**Distribution** Mediterranean, Atlantic, English Channel and North Sea. N.B. related species grow on rocks, etc.

### Zoobotryon verticillatum (Delle Chiaje)

Branched, tufted colony up to 500mm high; zooecia round or oval arranged spirally on branch tips.

Colour opaque-white, occasionally greenish.
Habitat growing in clusters on submerged objects, especially in harbours.
Distribution Mediterranean and adjacent Atlantic. N.B. this is an important
fouling organism.

### Alcyonidium gelatinosum (Linnaeus)

Gelatinous, smooth, sponge-
like colony up to 300mm long; individuals set into the mass.

Colour yellow-green-grey-brown.
Habitat growing on rocks and
shells from lower shore down
to about 100m.
Distribution Mediterranean,
Atlantic, English Channel,
North Sea and west Baltic.

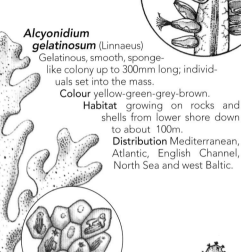

*Zoobotryon verticillatum*

*Alcyonidium gelatinosum*

### Alcyonidium polyoum (Hassall)

Similar to *A. gelatinosum* (above), but more irregular.
Habitat encrusting holdfasts of laminarians. N.B. not
found in the Mediterranean.

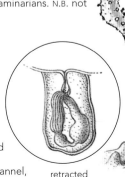

*Alcyonidium polyoum*

### Alcyonidium hirsutum

Flemming
Similar to *A. gelatinosum*
(above), but surface of colony
covered with small tubercles.
Habitat on rocky overhangs and
on seaweeds on middle and
lower shore.
Distribution Atlantic, English Channel,
North Sea and west Baltic.

retracted
zooid

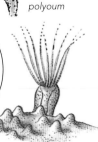

*Alcyonidium hirsutum*

245

## ORDER CHEILOSTOMATA

Zooecia generally flattened, and the zooecial walls are calcified. An anterior opening is present, with an operculum. Ooecia, avicularia and vildracula are present.

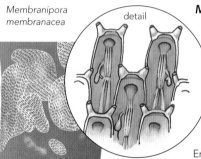

*Membranipora membranacea*

detail

### *Membranipora membranacea* (Linnaeus)
**Sea-mat**

Mat-like, encrusting colony of varying size, according to the substrate; round or irregular shape; zooids rectangular and beset with a blunt bristle on each side; elevated growths called 'towers' may occur.

**Habitat** on laminarians and other seaweeds from middle shore down to shallow water.

**Distribution** Mediterranean, North Sea, English Channel and Atlantic.

Whole colony on Laminaria frond

Whole colony on Laminaria frond

### *Electra pilosa* (Linnaeus)
**Hairy Sea-mat**

Similar to *Membranipora membranacea* (above). Very irregular and angular in outline; zooids bear 2 blunt bristles at one end, 1 conspicuous spine at the other end, and various smaller bristles all round.

**Colour** silver-grey.

**Habitat** on laminarians, other seaweeds and stone from middle shore down to shallow water.

**Distribution** Mediterranean, Atlantic, English Channel and North Sea.

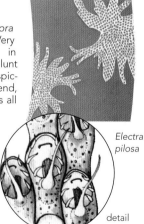

*Electra pilosa*

detail

*Electra crustulenta*

### *Electra crustulenta* (Pallas)

Similar to *Membranipora membranacea* (above). Zooids of varying shape but often pear-shaped with 1 spine at one end.

**Habitat** in shallow water often where salinity is low, encrusting stones, *Fucus*, *Zostera* (see pages 38–39, 64–65) and reeds.

**Distribution** Atlantic, English Channel, North Sea and Baltic.

### *Cellaria fistulosa* (Linnaeus)
Forms large growths up to 100mm high; jointed and dichoto-
mously branched; zooecium rounded to diamond shaped.
**Colour** ivory-white.
**Habitat** on hard and soft substrates, sometimes associated
with other bryozoans, from 20–200m.
**Distribution** Mediterranean, Atlantic, English
Channel and North Sea

### *Flustra foliacea* (Linnaeus)
**Hornwrack**
Leaf-like, branching colony up to 200mm
high; almost rectangular zooids on both
sides of colony; usually 2 short bristles on
either side of zooecium at one end.
**Colour** brown-green-yellow-grey; colour may face if
washed up.
**Habitat** on rocks and stones where it
may form extensive communities
providing food and shelter for a
variety of organisms; from shallow
water down to 100m.
**Distribution** Mediterranean,
Atlantic, English Channel and
North Sea.

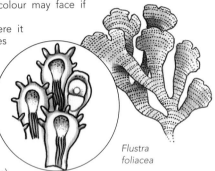

*Cellaria
fistulosa*

*Flustra
foliacea*

### *Securiflustra securifrons*
(Pallas)
Very similar to *Flustra foliacea* (above).
Reaching up to 150mm high; colonies are
divided into narrower 'leaves' and the zooecia are
more oblong than those of *Flustra foliacea*.
**Colour** pale yellowish.
**Habitat** as for *Flustra foliacea* (above).
**Distribution** some records from the
Mediterranean, but generally Atlantic
north of Wales, and North Sea.

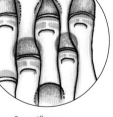

*Securiflustra
securifrons*

### *Bugula turbinata* Alder
Tufted, branching colonies up to 50mm high;
under a hand lens the spiral arrangement
of the branches, small elongated zooids with 2 bristles
at one end and the aviculariae may be seen.
**Colour** usually orange.
**Habitat** under overhanging rocks among
red seaweeds and sponges on the lower
shore and in shallow water.
**Distribution** Mediterranean, Atlantic and
English Channel.

*Bugula
turbinata*

247

### *Bugula neritina* (Linnaeus)

(Not illustrated) Similar to *B. turbinata* (page 247). Colonies up to 100mm high; stout and bushy; branching dichotomous; zooecia relatively large and rectangular; a short spine is present at the outer distal angle of each zooecium; large ooecia.
**Colour** brownish.
**Habitat** on lower shore and in deeper water.
**Distribution** Mediterranean, Atlantic, English Channel and North Sea.

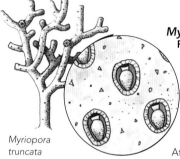

### *Myriopora truncata* (Pallas)
**False Coral**

Colony resembling a coral; branches have flat tips and lack the septa of true corals; reaching up to 100mm high; zooids embedded in the stems.
**Colour** yellow-red.
**Habitat** on rocks, in crevices and caves, usually in shallow, shaded places.
**Distribution** Mediterranean and adjacent Atlantic.

*Myriopora truncata*

### *Pentapora fascialis* (Pallas) *( =Hippodiplosia fascialis)*

Large and conspicuous colonies reaching 200mm high; oval- to rhomboidal-shaped zooids tightly grouped.
**Colour** orange-pink.
**Habitat** on hard substrates, among corals, etc., down to depths of about 25m.
**Distribution** Mediterranean (especially Adriatic). N.B. in British waters the somewhat similar *P. foliacea* (Ellis & Solander) (not illustrated) is usually found.

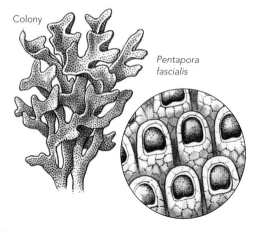

Colony

*Pentapora fascialis*

### *Margaretta cereoides* (Ellis & Solander) ( *=Tubucellaria opuntioides)*

Branching colony up to 50mm high; oval zooecia pressed together forming part of the stem.

**Colour** yellow-brown.

**Habitat** in shallow water among *Zostera* and *Posidonia* (see pages 64–65) roots.

**Distribution** Mediterranean.

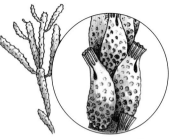

*Margaretta cereoides*

### *Pentapora foliacea* (Ellis & Solander)
**Ross Coral**

Large and conspicuous erect calcified colonies of sheet-like calcified fronds which join together.

**Colour** brown-orange up to 1m across.

**Habitat** on shells and rocks in shallow and deeper water.

**Distribution** Atlantic north to south-west Britain.

*Pentapora foliacea*

# PHYLUM ECHINODERMATA

The Echinodermata constitute a most distinct phylum in the animal kingdom and have been regarded by some authorities as being related to the ancestors of the chordates. Echinoderms are exclusively marine, and they display in their adult form a unique type of symmetry. This is essentially radial, with a mouth in the centre on one side of the body, and the anus normally in the centre on the opposite side. The body may be disc-shaped or globular, as in the sea-urchins, or it may be drawn out into five or more radii, as in the starfishes and brittle-stars. This symmetry is known as *pentamerism*.

Present-day echinoderms are divided into five distinct classes, although fossil evidence shows that other groups existed. Each of these extant classes is represented in the European area: they are the Crinoidea (feather-stars); the Asteroidea (starfishes); the Ophiuroidea (brittle-stars); the Echinoidea (sea-urchins, sand-dollars and heart-urchins); and the Holothuroidea (sea-cucumbers). All of these animals have a number of features in common, which usually make it easy to recognize echinoderms as such, yet at the same time they have features sufficiently different to allow one to decide which group they represent.

Echinoderms are triploblasts (i.e. their bodies develop from three embryonic cell layers: ectoderm, mesoderm and endoderm). The skeleton is basically internal, but it some-times protrudes to the exterior as in the spines of a sea-urchin. The skeleton consists of many plates of calcium carbonate; some of these are linked together to form the *test* or so-called 'shell', while others are mounted on the test to form spines. In most species apart from the echinoids the test plates are loosely connected together so that the animals are relatively flexible. The outer surface of the test is covered by a thin layer of epidermal cells which are often highly pigmented. Most of the organ systems lie within the large body cavity carried inside the test. These include the digestive and reproductive systems as well as the greater part of the unique water-vascular system. There appears to be no distinct osmoregulatory system, and perhaps this is why echinoderms cannot tolerate reductions in salinity. The sexes are usually separate and synchronous spawning often takes place at certain times of the year. External fertilization occurs in the sea and leads to the formation of a pelagic larva which passes through several stages in the plankton before settling to metamorphose into a juvenile.

The water-vascular system is apparent on the outside of the test in the form of double rows of *tube-feet*. These rows of tube-feet are known as ambulacra, and there are usually five (one to each radius of the body). Each tube-foot is elastic and can extend by being filled with fluid under pressure from within the body. Muscles along the tube-foot shaft can contract, and empty the fluid, and these muscles can also cause the tube-feet to bend, thus allowing locomotory 'steps'. Most starfishes, as well as the sea-urchins and sea-cucumbers, have suckers on their tube-feet, and

▼ External features and symmetry of the starfish *Astropecten irregularis*

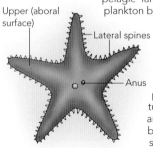

Upper (aboral surface)

Lateral spines

Anus

these can grip the substrate and serve as locomotory organs. This does not occur so much in the feather-stars and brittle-stars which move more by flexing their rays or arms. The internal anatomy of the water-vascular system is complex, and although it appears to open to the exterior by a special sieve-like test plate (the madreporite) there is little evidence for water movements in and out of it. In addition to locomotion, the water-vascular system is involved with respiration and feeding.

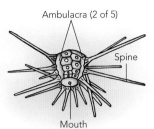

▲ External features of the sea-urchin *Cidaris cidaris* (some spines removed)

In the crinoids the mouth and anus are borne on the same side of the disc and face away from the substrate. Crinoids anchor themselves to the substrate by means of special appendages (cirri) on their under-sides, and use their tube-feet, which lack suckers and which occur in great numbers on the branching arms, to filter the sea water and collect from it small particles of suspended food matter. These particles are then passed down the arms to the mouth.

Asteroids hunt their prey by relying on a sense of smell to track it down. They frequently evert their stomachs over the prey or insert it inside gaping shells, etc. Digestion then occurs and the products can be absorbed. The mouth is on the underside and is thus well positioned for such behaviour. The small spines of asteroids are not as freely movable as those of the echinoids.

Ophiuroids are scavengers or particle feeders and use their suckerless tube-feet to collect food and pass it to their mouths. They move by means of their flexible arms. An anus is lacking and waste is passed out through the mouth.

Most round sea-urchins live by browsing on plant and animal growths which cover the rocks. A five-toothed chewing organ (the lantern of Aristotle) is carried inside the test, which also bears movable spines which serve a variety of functions including locomotion, defence and sensory detection. In some of these functions the spines are assisted by minute pincer-like organs (pedicellariae) which help to keep the test surface clean, and which also assist with defence. The tube-feet have suckers and can usually be extended considerable distances when they often resemble guy-ropes as they assist in posture and locomotion. The sand-dollars and heart-urchins are modified for burrowing and derive their food from organic substances in the sand and gravel where they live.

The skeletal elements of the holothuroids are often consid-erably reduced, giving the animals a softer texture. Modified tube-feet are arranged around the mouth to gather food either by filtering sea water or by sweeping organic matter from the surface of sand. The sea-cucumbers usually progress with one end leading, in the same manner as the sand-dollars and heart-urchins. In some cases (e.g. *Synapta*), the remaining tube-feet are not developed and the animal moves in a worm-like fashion. Picton, B. 1993, Mortensen, T. 1927 and Southward, E. and Campbell, A. 2005 provide further accounts of most European echinoderms. Nichols, D. 1969 gives a very good account of the biology of echinoderms.

## CLASS CRINOIDEA Feather-stars

Echinoderms with 5 paired arms bearing branches and growing from a cup-shaped body. These arms, with their tube-feet, are used for filtering food from sea water and for creeping and swimming. The mouth and anus are situated on the upper side and below the body are a number of short appendages (cirri) which grip the substrate.

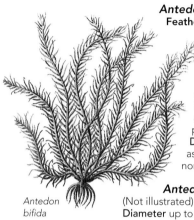

### *Antedon bifida* (Pennant)
**Feather-star**

**Diameter** up to 150mm or more overall.

**Body** bears inconspicuous disc with arms which often wave slowly in the water, outer surface of arms is not smooth.

**Colour** red-brown; arms sometimes banded with white.

**Habitat** on rocks and in crevices, sometimes attached to other growths, on lower shore in pools and down to about 200m.

**Distribution** Atlantic north from Portugal and as far as the Shetlands, English Channel and northern North Sea.

*Antedon bifida*

### *Antedon petasus* (Düben & Koran)
(Not illustrated)

**Diameter** up to 260mm overall.

**Body** appears more symmetrical than *A. bifida* (above) and the outer surface of the arms is smooth.

**Colour** arms red or white and brown.

**Habitat** on rocks, stones and seaweeds when growing on soft substrates, as well as on gorgonians, down to 200m.

**Distribution** Atlantic from south-west Britain and Ireland to Scandinavia.

## CLASS ASTEROIDEA Starfishes

Echinoderms in which the body is drawn out into distinct arms or rays, generally 5 in number. The mouth is on the underside and an anus on the upper side. Locomotory organs (tube-feet) are carried on the underside of each ray, and are generally armed with suckers. Clarke, A. M. and Downey, M. E. 1992 give more details on asteroids.

### *Luidia ciliaris* (Philippi)

**Diameter** up to 400mm or more overall.

**Body** always has 7 flattened rays; tube-feet end in knobs rather than suckers.

**Colour** orange-red on upper side; white below.

**Habitat** on sand and mud, sometimes buried, from extreme lower shore (occasionally) down to 150m.

*Luidia ciliaris*

**Distribution** Mediterranean, Atlantic, English Channel and North Sea.

**Luidia sarsi** Düben & Koren
(Not illustrated)
**Diameter** up to 200mm overall.
**Body** similar in shape to *L. ciliaris* (opposite), but never having more than 5 rays.
**Colour** yellow-red-brown above, with ray sides often darker.
**Habitat** on soft substrates from 10m downward.
**Distribution** Mediterranean, Atlantic, English Channel and northern North Sea.

*Astropecten
irregularis*

**Astropecten irregularis** (Pennant)
**Diameter** up to 120mm overall.
**Body** flattened, with very distinct 5-rayed appearance; each ray is edged with 2 layers of distinct marginal plates; each upper plate bearing 1 or 2 small spines (visible when viewed from above), lower layer has distinctly longer spines; tube-feet lack true suckers.
**Colour** orange-brown above; white below.
**Habitat** on sandy substrates often burrowing from extreme lower shore downward.
**Distribution** western Mediterranean, Atlantic, English Channel, North Sea and west Baltic.
N.B. the polychaete *Acholoë astericola* (see page 117) may be living between the tube-feet.

**Astropecten
aranciacus** (Linnaeus)
**Diameter** up to 600mm overall.
**Body** similar in shape to *A. irregularis* (above); each upper edge plate bears 2 or 3 strong, conical spines (visible when viewed from above).
**Colour** brown with reddish marks above; pale below.
**Habitat** as for *A. irregularis*.
**Distribution** Mediterranean and Atlantic north to Portugal. N.B. for a key to other species of *Astropecten* see Koehler, R. 1921.

*Astropecten
aranciacus*

**Porania pulvillus** (O. F. Müller)
**Diameter** up to 100mm over all.
**Body** fleshy and cushion-like with relatively short rays; smooth and rather sticky to the touch.
**Colour** often brilliant scarlet or orange-white above; white below.
**Habitat** on gravel from 10–250m, *P. pulvillus* feeds on falling organic particles.
**Distribution** Atlantic, English Channel and northern North Sea.

*Porania
pulvillus*

253

### Ceramaster placenta
(J. Müller & Troschel)
**Diameter** up to 160mm overall.
**Body** flat, pentagonal and very solid.
**Colour** brown-yellow-red.
**Habitat** on soft substrates from about 30m downward.
**Distribution** Mediterranean and Atlantic north to Biscay.

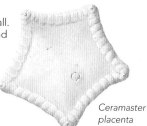

*Ceramaster placenta*

### Ceramaster granularis (O. F. Müller)
(Not illustrated)
**Diameter** up to 80mm overall.
**Body** pentagonal and similar to *C. placenta* (above), though more pointed.
**Habitat** from 20–1400m.
**Distribution** Atlantic from Morocco to Greenland, North Sea and Skagerrak.

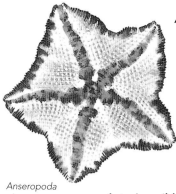

### Anseropoda placenta (Pennant)
### ( =Palmipes membranaceus)
**Goose-foot Star**
**Diameter** up to about 150mm overall.
**Body** very flat and pentagonal; slightly concave edges which often look rather tattered.
**Colour** brilliant, often with patterns of red and white above; yellow-white below (detailed distribution of pigment is variable).
**Habitat** on sand and mud from 10–100m.
**Distribution** Mediterranean, Atlantic, English Channel and northern North Sea.

*Anseropoda placenta*

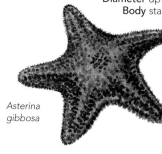

### Asterina gibbosa (Pennant)
**Cushion-star** or **Starlet**
**Diameter** up to 50mm overall, though occasionally more.
**Body** star-shaped with rounded tips to the rays; not as flat as *Anseropoda placenta* (above).
**Colour** green-pale brown on upper surface; more yellow below.
**Habitat** on and under rocks and stones on the lower shore and down to 100m.
**Distribution** Mediterranean, Atlantic and English Channel. N.B. possibly more tolerant to fresh water than most starfishes.

*Asterina gibbosa*

### *Asterina phylactica* Emson and Crump
**Diameter** up to 15mm.
**Body** 5-sided with rounded tips to the rays, tube feet suckered, Body not as flat as *Anseropoda* (opposite).
**Colour** greenish with a dark-brownish red over the radii.
**Habitat** in rocks pools and also in shallow water.
**Distribution** Mediterranean and Atlantic north to south-west Britain.

*Asterina phylactica*

### *Echinaster sepositus* Gray
**Diameter** up to 200mm overall.
**Body** small disc covered with soft skin; fairly conspicuous pockmarks on upper surface and larger disc distinguish it from *Ophidiaster ophidianus* (not illustrated); relatively long rays taper gradually; tube-feet with suckers.
**Colour** scarlet.
**Habitat** on rocks and softer substrates from 1–250m.
**Distribution** Mediterranean and Atlantic north to Brittany.

### *Henricia oculata* (Pennant)
(Not illustrated) Similar to *H. sanguinolenta* (below).
**Diameter** up to 140mm overall.
**Body** relatively small disc with stiff rays which are nearly circular in section and continuously tapering. Ornamented with small but quite stout spines with rounded tips bearing numerous short thorns.
**Colour** red, orange or yellow above, whiter below.
**Habitat** on soft substrates and among pebbles and small stones.
**Distribution** Atlantic from Biscay northward, English Channel, northern North Sea and west Baltic.

*Echinaster sepositus*

### *Henricia sanguinolenta* (O.F. Müller)
**Diameter** up to 200mm thogh often less.
**Body** relatively small disc ornamented with minute spines each carrying 5–7 thorns at their tips, and with stiff arms which are nearly circular in section and tapering along their length to narrow tips.
**Colour** blood-red to purple above, white below.
**Habitat** lower shore and shallow water down to deep water.
**Distribution** Atlantic north to Scandinavia and Baltic.

*Henricia sanguinolenta*

### *Crossaster papposus* (Linnaeus) *( =Solaster papposus)*
**Common Sun-star**

Diameter up to 250mm overall, but frequently smaller.
Body bears large disc with between 8–13 blunt rays which are rarely longer than half the diameter of the disc; the whole surface of the animal is covered with small but distinct spines.

Colour variable; brown-red with white markings above, yellow-white below; *C. papposus* is often beautifully patterned.

Habitat on sand and stones and among mussel and oyster beds in the company of other starfishes from 10–40m; frequently preys upon other echinoderms.

Distribution Atlantic from France northwards and to Greenland, English Channel, North Sea and west Baltic.

*Crossaster papposus*

### *Solaster endeca* (Linnaeus)
**Purple Sun-star**

Diameter up to 300mm overall, but frequently smaller.
Body 7–13 rays; spines not as apparent as those of *Crossaster papposus* (above); upper surface has a fairly hard texture.

Colour purple-orange above; white-orange colour below.

Habitat on hard substrates from 20–90m.

Distribution west and north Atlantic coasts of the British Isles and north to Greenland, North Sea and west Baltic.

### *Asterias rubens* Linnaeus
**Common Starfish**

Diameter may reach up to 500mm, but frequently much smaller.

*Solaster endeca*

Body has plump, rounded, tapering rays often slightly turned up at the tip when active; surface covered with irregularly arranged spines surrounded by pedicellariae (see *Marthasterias glacialis* right); tube-feet bear suckers.

Colour brown-yellow above; paler below.

Habitat on rocks and stony ground, in mussel and oyster beds from the lower shore down to avout 200m; sometimes gregarious.

Distribution Atlantic, English Channel, North Sea and west Baltic.

*Asterias rubens*

## *Marthasterias glacialis* (Linnaeus)
**Spiny Starfish**
**Diameter** may reach up to 800mm overall, but frequently much smaller.
**Body** rounded, gradually tapering rays often slightly turned up at tips when active; body surface is covered in conspicuous spines surrounded by rings of small, pincer-like organs (pedicellariae) which can easily be discerned with a hand lens; tube-feet bear suckers.
**Colour** brown-yellow with green-grey markings above, white-yellow below.
**Habitat** on rocks and stony substrates from the lower shore down to 180m.
**Distribution** Mediterranean, Atlantic, western English Channel and North Sea.

*Marthasterias glacialis*

## *Leptasterias mülleri* (M. Sars)
(Not illustrated)
**Diameter** up to 100mm overall.
**Body** somewhat similar in shape to *Asterias,* but with stouter dorsal skeleton due to crowding together of the plates; upper surface often with many knobby spines arranged in rows; pedicellariae present.
**Colour** disc and inner part of rays red-violet above.
**Habitat** among rocks from middle shore downward.
**Distribution** Atlantic north of English Channel, and North Sea.

## *Coscinasterias tenuispina* (Lamarck)
**Diameter** up to 150mm overall.
**Body** has relatively small disc, bearing 6–10 rays often of different lengths; surface covered with distinct spines each surrounded by pedicellariae (see *Marthasterias glacialis* above).
**Colour** variable; basic shade may be white, red-brown or purple, with blue or brown spots.
**Habitat** on rocks and stones from the lower shore down to 30m.
**Distribution** Mediterranean and Atlantic, north to Portugal.

*Coscinasterias tenuispina*

## *Stichastrella rosea* (O. F. Müller)
**Diameter** up to 150mm overall.
**Body** small disc bearing long, tapering rays of equal length; surface covered with small spines in groups; pedicellariae present.
**Colour** generally red-orange-yellow.
**Habitat** on hard and soft substrates from 4–350m.
**Distribution** Atlantic north from Biscay to Norway, Irish Sea and English Channel west to Devon.

*Stichastrella rosea*

section of
arm of
*O. fragilis*

# CLASS OPHIUROIDEA Brittle-stars

Echinoderms normally with 5 unbranched, jointed arms bearing spines; the tube-feet on the underside of the arms lack suckers. The rounded disc is flattened, and the mouth is on the underside. There is no anus nor are there pedicellariae. The precise identification of brittle-stars is difficult unless account is taken of small structures best seen through a hand lens. A useful identification feature in the ophiuroids is the arm spine number (a.s.n.), which is the number of spines on one side only of one arm joint. Mortensen, T. 1927 and Southward, E. C. and Campbell, A. C. 2005 provide good accounts of many European ophiuroids.

### *Ophiothrix fragilis* (Abildgaard)
**Common Brittle-star**
Disc up to 20mm in diameter, often pentagonal with the points visible between the arms; upper surface has many minute spinelets, and some longer spinelets, arranged in 5 radiating, v-shaped groups over each arm and on either side of 2 naked, triangular plates.

*Ophiothrix fragilis*

Arms not more than 5 times disc diameter; fragile and often broken or regenerating; arm spines conspicuous and finely thorned; lowermost arm spine hooked; a.s.n. = 7.
Colour variable; bright red-brown-violet-purple or patterned above.
Habitat under stones, seaweeds and shells from lower shore down to 350m.
Distribution Mediterranean, Atlantic, English Channel and North Sea.

### *Ophiocomina nigra* (Abildgaard)
Disc up to 30mm in diameter, faintly pentagonal with arms borne at points, upper surface covered with fine granules.
Arms tapering, and not more than 5 times disc diameter; arm spines, fine and glassy and equal to 2 arm joints in length; lower arm spines shorter; a.s.n. = 5–7.
Colour black-brownish grey.
Habitat among rocks, seaweed and sand, often shaded, from lower shore down to 400m.
Distribution Mediterranean, Atlantic, English Channel and North Sea.

### *Amphiura brachiata* (Montagu)
Disc up to 10mm in diameter. Arms exceptionally long, may reach 15 times disc diameter; a.s.n. = 8–10 on joints near disc.
Colour grey-brown.
Habitat burrowing in sand with arms looped, coiled or twisted, from lower shore down to 40m.

*Ophiocomina nigra*

Distribution Atlantic north from Biscay, English Channel and North Sea. N.B. a related species, *A. neapolitana* (M. Sars) (not illustrated) occurs in the Mediterranean. A

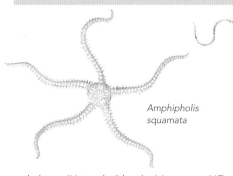

Amphipholis
squamata

Amphiura
brachiata

polychaete (*Harmothoë lunulata*) (see page 117)
and a bivalve (*Montacuta bidentata*) may
be associated with this species.

### Amphipholis squamata (Delle
Chiaje)
**Disc** up to 5mm in diameter with 2 conspicuous pale
plates above each arm.
**Arms** up to 4 times disc diameter; arm spines short and conical;
a.s.n. = 4, on joints near the disc.
**Colour** bluish-grey-white.
**Habitat** under rocks, pebbles and seaweeds (especially
corallines) from lower shore down to 250m.
**Distribution** Mediterranean, Atlantic, English Channel and
North Sea. N.B. often present in great numbers.

### Ophiura texturata Lamarck
**Disc** up to about 30mm, rounded from above and
scaled; 2 conspicuous plates above the origin of
each arm.
**Arms** tapering, up to 4 times disc diameter;
arm spines tapering, shorter than the
arm width at that point and lying
against the arm itself; a.s.n. = 3.
**Colour** orange-brown above; pale below.
**Habitat** burrowing in sand from lower shore down to 200m
(may get washed to upper shore).
**Distribution** Mediterranean, Atlantic, English Channel, North
Sea and west Baltic. N.B. a closely related species *O. albida*
(not illustrated) may befound in similar habitats (see
Koehler, R. 1921 and Mortensen, T. 1927 for
differences).

Ophiura
texturata

### Amphiura filiformis O.F. Müller
**Disc** up to 10mm in diameter, two conspicuous
plates above the origins of each arm.
**Arms** up to a.s.n 5–7.
**Colour** red-grey.
**Habitat** in shallow water on sand and mud.
**Distribution** Mediterranean, Atlantic north to
Norway, English Channel and North Sea.

Amphiura
filiformis

259

# Class Echinoidea
## Sea-urchins, sand-dollars and heart-urchins
### Subclass Perischoechinoidea
### Order Cidaroida and Subclass Euechinoidea
### Superorders Diadematacea and Echinacea

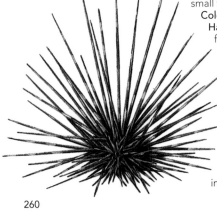

*Cidaris*
*cidaris*

Spherical echinoderms with a chalky, shell-like skeleton (the test) bearing mobile spines mounted externally on small knobs and being perforated by 5 double rows of pores which in life permit fluid to pass into the long tube-feet. The mouth is on the underside and has chewing teeth, and the anus is at the apex of the test. See Southward, E. and Campbell, A. C. 2005 for further details.

### Cidaris cidaris (Linnaeus) ( =Dorocidaris papillata)
**Test** up to 70mm in diameter.

**Spines** both large and small; the large spines have longitudinal ridges (made up of rows of very fine thorns) reaching twice the test diameter and often encrusted with sponges, hydroids, etc.; the small spines are arranged around the bases of the large spines and on either side of the rows of tube-feet.

**Colour** yellow-green-red-brown.

**Habitat** various; from 30m downward, test occasionally washed up after storms especially on south-west coasts.

**Distribution** Mediterranean, Atlantic north to Biscay and in deep water off the west coast of Ireland.

aboral view
of test

### Stylocidaris affinis (Philippi)
(Not illustrated) Similar to *Cidaris cidaris* (above) but smaller.

**Test** up to 40mm in diameter.

*Centrostephanus*
*longispinus*

**Spines** large and tapering; usually reach a little more than the test diameter and bear many visible small thorns.

**Colour** orange-brown.

**Habitat** on coralline seaweeds and rocks from the lower shore down to 30m.

**Distribution** Mediterranean and Atlantic north to Portugal; not found in the Adriatic.

### Centrostephanus longispinus (Philippi)
**Test** up to 60mm in diameter.

**Spines** long, hollow, slender, mobile.

**Colour** spines patterned with brown and white bands; test red-brown.

**Habitat** normally below 40m and often in very deep water.

Distribution Mediterranean and adjacent Atlantic. N.B. the spines of this urchin may inflict painful wounds.

*Arbacia lixula*

### *Arbacia lixula* (Linnaeus)
**Black Sea-urchin**
Test up to 50mm in diameter.
Spines up to 30mm long; solid and with sharp tips.
Colour black spines; the cleaned test is pink with characteristic red lines marking the position of the tube-feet pores and the very large oral opening.
Habitat on rocks and among coralline seaweeds from the extreme lower shore down to 40m.
Distribution Mediterranean and Atlantic coasts to Portugal. N.B. this urchin is often confused with *Paracentrotus lividus* (see page 262) but in life the two may easily be distinguished by the extent of the soft membranes overlying the oral opening of the test. This opening is large in *A. lixula* and consequently the membrane is extensive; in *P. lividus* it is smaller and the membrane less apparent.

oral view
of test

aboral
view of
test

### *Sphaerechinus granularis* (Lamarck)
**Violet** or **Purple Sea-urchin**
Test up to 120mm in diameter.
Spines up to 20mm long; short and solid.
Colour this urchin may easily be recognized by its colours, the spines often having white tips and purple shafts, or being entirely white and conspicuous against the purple of the test; the cleaned test can be recognized by the 10 narrow slits (each about 2mm long) which occur around the oral opening or the underside.
Habitat on rocks and coralline seaweeds from the extreme lower shore down to 100m.
Distribution Mediterranean and Atlantic north to Channel Islands.

*Sphaerechinus granularis*

oral view
of test

### *Strongylocentrotus droebachiensis*
(O. F. Müller)
Test up to 50mm in diameter.
Spines up to 20mm long; solid.
Colour greenish.
Habitat on rocks and seaweeds from sea level down to 1200m.
Distribution north Atlantic coasts of Shetland and Norway, and North Sea. N.B. this urchin cannot be easily confused with *Paracentrotus lividus* (see page 262) because their distributions do not normally overlap.

*Strongylocentrotus droebachiensis*

261

### *Paracentrotus lividus* (Lamarck)

*Paracentrotus lividus*

*Paracentrotus lividus*

whole animal

test oral view

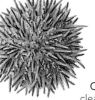

**Test** up to 60mm in diameter.

**Spines** up to 30mm long; smooth and solid.

**Colour** variable; from green to dark brown; test when cleaned shows a relatively small oral opening (c.f. *Arbacia lixula*, page 261).

**Habitat** on rocks and stones and among coralline seaweeds from lower middle shore in rock pools down to 30m.

**Distribution** Mediterranean and Atlantic north to Channel Islands and the west coast of Ireland; very rarely in the western English Channel. N.B. this species sometimes bores into the rocks of the shore, especially those on many western Irish beaches; it often covers itself with small fragments of seaweeds and shells, etc.; this species is somewhat gregarious and may occur in flocks, and in the Mediterranean it may be found associated with *Arbacia lixula*. In France and elsewhere the roes of this species are considered a delicacy.

*Psammechinus miliaris*

whole animal

test aboral view

### *Psammechinus miliaris* (Gmelin)
**Green Sea-urchin**

**Test** up to 40mm in diameter and very occasionally more.

**Spines** up to 15mm long; rather coarse.

**Colour** spines have violet tips; test green when cleaned.

**Habitat** on rocks and under stones often associated with coralline seaweeds and other encrusting organisms, from the lower middle shore down to 100m.

**Distribution** Atlantic, English Channel, North Sea and west Baltic.

### *Psammechinus microtuberculatus* (Blainville)
(Not illustrated)

**Test** up to 35mm in diameter.

**Spines** up to 15mm long; slender.

**Colour** spines have reddish tips; test when cleaned is green-grey.

**Habitat** on rocks and stones from 4–100m.

**Distribution** Mediterranean and Atlantic north to Portugal. N.B. this species could easily be confused with *P. miliaris* (above) but for the fact that their distributions hardly ever overlap. it is also more delicate in form.

## *Echinus esculentus* Linnaeus
**Edible Sea-urchin**

**Test** frequently up to 100mm in diameter, but may reach 170mm.

**Spines** about 15mm long; short and solid; not clearly divided into different size groupings (primaries and secondaries) and quite abundant on the test.

**Colour** spines often have purple tips; test when cleaned is beautiful, with shades of red and purple, with white spine attachment points.

**Habitat** on rocks and among seaweeds from extreme lower shore down to 50m.

**Distribution** Atlantic from Portugal to Norway, English Channel and North Sea. N.B. the polychaete *Flabelligera affinis* and the amphipod *Astacilla longicornis* (see pages 129 and 213) may be found living among the spines. This species has been used for food in various countries in Europe including Britain and Portugal, but the roes are rather coarse in comparison with those of *Paracentrotus lividus* (opposite).

whole animal

*Echinus esculentus*

## *Echinus acutus* Lamarck
Similar to *E. esculentus* (above).

**Test** up to 160mm in diameter; noticeably more conical and not so well covered with spines which are rather scarce on the upper regions.

**Spines** more readily divided into two size groupings (primaries and secondaries); longer than those of *E. esculentus*.

**Colour** more reddish than in *E. esculentus* and often with white points.

**Habitat** often on soft substrates from 20–1000m.

**Distribution** Mediterranean (but not in the Adriatic), Atlantic, English Channel (rare) and the northern North Sea.

test lateral view

*Echinus acutus*

## *Echinus melo* Lamarck
(Not illustrated)

**Test** up to 170mm in diameter; there is some discussion as to the validity of this species which is very similar in many ways to *E. acutus* (above), but the test is globular rather than conical.

**Colour** brownish red.

**Habitat** on rocky substrates.

**Distribution** Mediterranean (including the Adriatic) and Atlantic north to Portugal.

test aboral view

*Echinus acutus*

test lateral view

## Superorder Gnathostomata
## Order Clypeasteroida
### Sand-dollars

Bilaterally symmetrical, disc-shaped echinoids with tube-feet which are mainly restricted to the upper side and are often arranged in petal-like patterns. The anus is on the underside, sometimes close to the mouth which bears chewing teeth.

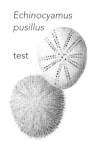

*Echinocyamus pusillus*

test

### *Echinocyamus pusillus* (O. F. Müller)
**Pea-urchin**
**Test** up to 15mm long.
**Spines** short and thickly set.
**Colour** green-grey.
**Habitat** in sand and gravel from 1–800m.
**Distribution** Mediterranean, Atlantic, English Channel, North Sea and west Baltic.

## Superorder Atelostomata
## Order Spatangoida
### *Heart-urchins*

Bilaterally symmetrical, heart-shaped echinoids with most tube-feet borne on the upper side but a few arranged below. The mouth and the anus are situated on the underside, towards the anterior and posterior ends respectively. The mouth lacks chewing teeth. The test is thickly covered with fine spines, often with a fur-like appearance. These urchins are highly adapted for burrowing.

*Spatangus purpureus*

test

test lateral view

### *Spatangus purpureus* O. F. Müller
**Purple Heart-urchin**
**Test** up to 120mm long, bearing five rows of tube-feet of which the anterior row is the longest and lies in a pronounced but relatively shallow furrow leading towards the mouth.
**Spines** mostly short, but some on the upper side are longer.
**Colour** red-violet in life, but grey-white when cleaned.
**Habitat** in coarse sand and shell gravel from 5–800m.
**Distribution** Mediterranean, Atlantic, English Channel and North Sea.

### *Echinocardium cordatum* (Pennant)
**Sea-potato**

**Test** up to 90mm long, but often smaller; bears 5 rows of tube-feet of which the anterior row is the longest, and appears modified considerably from the rest lying in a deep furrow reaching nearly to the mouth.
**Spines** mostly short, but some are long and curved; densely distributed and directed backward.
**Colour** yellow-brown in life, but yellow-white when cleaned.
**Habitat** in sand from the lower shore to 200m.
**Distribution** Mediterranean, Atlantic, English Channel and North Sea.

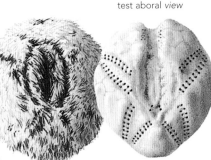

test aboral view

*Echinocardium cordatum*

### *Echinocardium pennatifidum* Norman
(Not illustrated)

**Test** up to 70mm long; bears 5 rows of tube-feet of which the anterior row is modified, being reduced and having relatively large pores; the frontal notch is inconspicuous and not severely furrowed as in *E. cordatum* (above).
**Colour** white-yellow in life; white when cleaned sometimes showing grey blotches.
**Habitat** in sand from 5–150m.
**Distribution** Mediterranean, Atlantic, English Channel and North Sea; not present in the Adriatic.

test oral view

### *Echinocardium flavescens* (O. F. Müller)
(Not illustrated) Very similar to *E. pennatifidum* (above) but in life has yellow-pink spines.
**Habitat** in sand from 20–325m.
**Distribution** Atlantic, English Channel, North Sea and west Baltic.

*Brissopsis lyrifera*

### *Brissopsis lyrifera* (Forbes)
**Lyre-urchin**

**Test** up to 70mm long; bears 5 rows of tube-feet, the posterior 2 rows being shorter and the anterior rows lying in a frontal notch which is not as deep as that of *Echinocardium cordatum* (above); the anus is fractionally above the edge of the test so as to appear slightly on the upper side.
**Spines** short, dense and fur-like.
**Colour** brown-red in life; yellow-grey when cleaned.
**Habitat** buried in sand from 5–300m.
**Distribution** Mediterranean, Atlantic and North Sea; not present in the Adriatic.

test

265

# Class Holothuroidea Sea-cucumbers

Bilateral echinoderms lacking conspicuous spines, arms or rays, and usually cucumber-shaped or worm-like. Tube-feet are present in many species, arranged in 5 rows along the sides of the animal; 3 rows are usually in contact with the substrate and these are equipped with suckers for locomotion, while the other 2 rows are borne away from the substrate. The anterior mouth is surrounded by modified tube-feet, and the anus is positioned posteriorly. A skeleton of loosely associated calcareous spicules is embedded in the 'skin'. Picton, B. 1993 gives excellent illustrations of these and other sea-cucumbers.

*Stichopus regalis* Cuvier

*Stichopus regalis*

**Length** up to 300mm.
**Body** has well-developed, flattened, sole-like underside bearing locomotory tube-feet; rounded upper side is covered with warty protuberances and bumps; mouth opens on the underside and is not quite at the front of the body.
**Colour** brownish with pale spots above; paler below.
**Habitat** on sandy substrates and among corals and bryozoans, from 5–400m.
**Distribution** Mediterranean and Atlantic.

*Holothuria forskali* Delle Chiaje
**Sea-cucumber** or **Cotton-spinner**

*Holothuria forskali*

**Length** up to 200mm.
**Body** cucumber-shaped; fairly well-defined lower surface bearing three rows of suckered, locomotory tube-feet; upper surface curved, warty and bearing irregularly arranged, suckerless tube-feet; around the mouth 20 feathery, modified tube-feet may be visible.
**Colour** black above; pale brown-yellow below.
**Habitat** on soft substrates and among *Zostera* (see page 64–65) from lower shore down to 70m.
**Distribution** Atlantic and English Channel.
N.B. if molested, this species may eject white sticky threads from the hind end as a defence mechanism.

*Aslia lefevrei* (Barrois)

*Aslia lefevrei*

**Length** up to 150mm, but often less.
**Body** generally more cylindrical than *Stichopus regalis* with lower surface not so clearly defined; well-developed locomotory tube-feet below arranged in double rows, upper tube-feet conspicuous but reduced; 10 special tube-feet around terminal mouth often extended and feather-like.
**Colour** leathery skin is dirty white to brown.
**Habitat** hidden under rocks and stones on the lower shore and down to 20m.
**Distribution** Atlantic and English Channel.

**Leptopentacta elongata** (Düben & Koren)
**Length** up to 150mm.
**Body** similar to *C. normani* (opposite); generally of curved shape and pointed towards the hind end; tube-feet apparently evenly distributed and arranged in 5 double rows.
**Colour** usually dark brown.
**Habitat** on muddy substrates from 5–150m.
**Distribution** Mediterranean, Atlantic north to Norway, English Channel and North Sea. N.B. several other species of *Cucumaria* occur (see Mortensen, 1927 and Koehler, 1921).

*Leptopentacta elongata*

**Thyone fusus** (O. F. Müller)
**Length** up to 200mm, but often less.
**Body** plump, tapering towards both ends; exterior covered by many irregularly arranged tube-feet; 10 modified oral tube-feet may be visible surrounding the mouth.
**Colour** variable, but generally white-pink.
**Habitat** on soft substrate, from 10–150m.
**Distribution** Mediterranean, Atlantic from Madeira to Norway and northern North Sea.

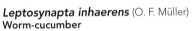

*Thyone fusus*

**Leptosynapta inhaerens** (O. F. Müller)
**Worm-cucumber**
**Length** up to 180mm.
**Body** has no tube-feet apart from 12 around the mouth modified for feeding, each of these carries 5–7 pairs of minute, finger-like branches; minute, anchor-shaped skeletal spicules used in locomotion protrude through the soft skin making it adhesive.
**Colour** usually pale pink.
**Habitat** on or burrowing in mud and sand from 10–50m.
**Distribution** Atlantic from French coast to Norway, English Channel and northern North Sea.

*Leptosynapta inhaerens*

**Labidoplax digitata** (Montagu)
Similar in some respects to *Leptosynapta inhaerens* (above).
**Length** up to 180mm, though occasionally more.
**Body** lacks locomotory tube-feet; oral tube-feet each have 2 pairs of minute, finger-like branches.
**Habitat** on mud and sand from lower shore down to 70m.
**Distribution** Mediterranean and Atlantic.

*Labidoplax digitata*

267

# Phylum Chaetognatha

The chaetognaths are a small group of exclusively marine animals. Not more than fifty species have been identified, and these are all placed in one class. Their bodies are difficult to see unless they are held up to the light in a jar of sea water. Although they will be rarely met with on the seashore, they will be frequently encountered in plankton samples. About twelve or so species have been recorded from European waters, but because they are small and usually oceanic, only two are mentioned here.

Certain species of chaetognath are typically associated with particular currents of ocean water, and because of this they are important to oceanographers who need to trace the origins of planktonic communities. Such species are known as *indicator species*, one of which is *Sagitta setosa,* a common indicator for coastal waters in the North Sea and English Channel regions.

The body plan of the chaetognaths is quite characteristic. Essentially a small, bilaterally symmetrical, torpedo-shaped body bears paired side-fins and a tail fin. The anterior mouth is equipped with strong, grasping spines. Circulatory and excretory systems are lacking. The main variations in the body plan are the number of side-fins and the general shape of the body.

Chaetognaths are active predators. Typical items in their diet are small crustaceans called copepods. Chaetognaths appear able to detect the presence of prey by the vibrations that are sent out when swimming. The prey is then captured with the help of the grasping spines around the mouth. More identification details for these animals are given by Newell, G. E. and Newell, R. C. 1973.

Sagitta
setosa

### *Sagitta setosa* J. Müller
**Arrow Worm**
**Length** up to 15mm.
**Body** narrow and transparent without colour; 2 pairs of lateral fins, the anterior being about halfway along the body, the posterior terminating a little in front of the tail.
**Habitat** planktonic in coastal waters, sometimes stranded in rock pools.
**Distribution** Mediterranean, Atlantic, English Channel, North Sea and west Baltic.

### *Spadella cephaloptera* (Busch)
**Length** up to 8mm.
**Body** less elongated than *Sagitta setosa,* but broader; 1 pair of lateral fins which are more or less continuous with the tail fin.

Spadella
cephaloptera

**Habitat** unlike the majority of chaetognaths, this is a benthic species which attaches itself to rocks and seaweeds by means of suckers. It may be found on the seabed or in rock pools on the lower shore.
**Distribution** Mediterranean, Atlantic and English Channel.

The hemichordates are another small group of exclusively marine animals. Until recently they were classified as a sub-phylum within the Chordata. About eighty species are known, and these are divided into three classes: Enteropneusta, Pterobranchia and Planctosphaeroidea. Of these, only the first class falls within the scope of this book.

Hemichordates possess bilaterally symmetrical, worm-like bodies divided into three distinct zones, with or without gill slits. A circulatory system is present, but an excretory system is lacking. Tentacles may be present on the middle zone of the body. Both solitary and colonial forms occur.

## CLASS ENTEROPNEUSTA Acorn worms

These solitary marine worms should be quite easily distinguished from the other worms described in this book by the manner in which their bodies are divided into three regions or zones: an anterior proboscis, a short collar lying immediately behind the proboscis, and a long abdomen. The relative shapes of the various parts of the body are important in classification. The mouth opens at the junction of the proboscis and collar, and gives rise to the gut which passes back through the abdomen. The abdomen itself may be rounded or flattened. At the front of the abdomen are gills which communicate with the gut and facilitate respiration. Tentacles are lacking. In some species the gut forms a number of pockets towards the rear of the body and these can be discerned as small lumps arranged on either side of the posterior part of the abdomen. They are known as *hepatic pouches*. There is a simple circulatory system and a simple nervous system.

*Saccoglossus cambrensis*

Enteropneusts live in u-shaped burrows in sand or mud, and may form characteristic casts reminiscent of polychaete worm casts. Like so many burrow-dwelling invertebrates, they are filter feeders extracting particulate food from sea water. The sexes are separate and the sperms and eggs escape into sea water by rupture of the body wall. A pelagic larva is formed after fertilization.

### *Saccoglossus cambrensis* Stiasny
**Length** up to 60mm.
**Body** proboscis very long, and red-deep pink in colour; abdomen round, slightly paler and lacking hepatic pouches (see introduction to this group).
**Habitat** burrowing in clean sand and gravel from lower shore downward.
**Distribution** Atlantic.

*Balanoglossus clavigerus*

### *Balanoglossus clavigerus* Delle Chiaje
**Length** up to 300mm.
**Body** proboscis short, and yellow in colour; abdomen flattened and pale brown.
**Habitat** burrowing in sand, mud and clay in shallow and deeper water.
**Distribution** Mediterranean, Atlantic and English Channel.

# PHYLUM CHORDATA

Chordates are animals possessing a single, hollow, dorsal nerve cord. A true body cavity (coelom) is present, as are gill slits and a notochord. The tail is post anal.

This phylum includes the fishes, which are the most familiar of all sea animals. A more detailed introduction is provided for them on pages 276–277. Far less familiar are the so-called invertebrate chordates (subphylum Urochordata), i.e. those animals whose evolutionary standing places them near the vertebrates, yet which as adults lack any trace of the diagnostic notochord or backbone, and which therefore resemble superficially the invertebrates. Two urochordate classes are treated here.

The class Thaliacea includes a number of organisms commonly known as salps. The adults lead a pelagic life, swimming amid their food supply (smaller planktonic organisms) which they filter by means of their gills. The body is surrounded by a number of muscle bands which can contract to force water out of the posterior exhalent opening, thus moving the animal forward. Water with food suspended in it is drawn in through the inhalent opening at the anterior end. The body structure is quite complicated, and a good deal of it is associated with the reproductive process. Thaliaceans have complex life cycles, and a very small tadpole-like larva with a notochord is formed. Some species are capable of forming associations or colonies. In some cases a species may be either colonial or solitary, and thus exists in two forms. The left-hand diagram illustrates the basic characters of a solitary salp.

The class Ascidiacea comprises the sea-squirts. Unlike the salps, these are bottom dwellers as adults, and live attached to rocks or other organisms. They do retain a small, free-swimming tadpole-like larva, however. They also filter suspended food particles from the water which they take into their gills for respiration. The body is usually encased in a thick tunic made from cellulose-like material which often has a jelly-like consistency. As with the salps, there are inhalent and exhalent openings, but these are generally both situated towards the

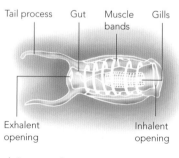

Tail process  Gut  Muscle bands  Gills

Exhalent opening    Inhalent opening

▲ Features of a solitary salp

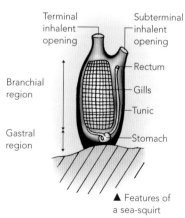

Terminal inhalent opening    Subterminal inhalent opening

Branchial region

Rectum

Gills

Tunic

Gastral region

Stomach

▲ Features of a sea-squirt

end opposite to the attachment point of the animal. The precise relationship of the two openings (the inhalent opening is normally terminal) is important in the identification of these animals. Another important feature is the proportion of body length which is occupied by the gill or branchial region, and the proportion which is occupied by the gastric region. In some species these areas are easier to discern than in others. Below the gills lie the stomach and reproductive systems as well as much of the circulatory system, but for simplicity the latter two systems have been omitted from the included diagram (right-hand diagram, opposite). Some sea-squirts, e.g. *Botryllus,* are colonial. Here, the individuals are arranged around common exhalent canals and are supported in a massive tunic.

# Subphylum Urochordata
## Class Thaliacea

Pelagic adults resemble floating, jelly-like barrels. A hollow, dorsal nerve cord and tail are present only in the larva. The nervous system is reduced. More identification details for these animals are given by Newell, G. E. and Newell, R. C. 1973.

### *Salpa maxima* Forskål
Solitary.
**Length** up to 100mm long.
**Body** barrel-shaped, with 9 conspicuous muscle bands arranged along the body.
**Habitat** planktonic.
**Distribution** Mediterranean and adjacent waters.

### *Salpa democratica* Forskål
**Length** solitary individuals up to 15mm; colonial individuals up to 6mm; colonial groups often have long streamers up to 300mm or more.
**Habitat** planktonic.
**Distribution** Mediterranean and adjacent waters.

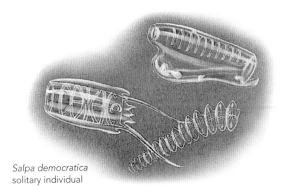

*Salpa maxima*

*Salpa democratica*
solitary individual

*Salpa democratica*
chain of colonial
individuals

# CLASS ASCIDIACEA

*Clavelina*
*lepadiformis*

A hollow, dorsal nerve chord is present only in the larva, which resembles a small tadpole. Adults are sessile or pelagic, hermaphrodite, and covered with a tunic of cellulose. They may be solitary or colonial. Berrill, N. J. 1950, and Millar, R. H. 1970, provide a more detailed account of many European ascidians.

### *Clavelina lepadiformis* (O. F. Müller)

Sessile and colonial; individuals reach 20mm in height and are easily distinguished from each other by being joined at the base only by thin stolons; inhalent and exhalent openings close together; branchial region markedly shorter than gastric region.

**Colour** transparent and jelly-like, with pink-yellow-white marks.

**Habitat** on stones, seaweeds and shells from extreme lower shore down to about 50m.

**Distribution** Mediterranean, Atlantic, English Channel and North Sea.

*Aplidium*
*proliferum*

part of colony with
individuals grouped
around common
exhalent opening

### *Aplidium proliferum* (Milne-Edwards)

Sessile; individuals growing in fleshy colonies forming club-shaped growths; these may reach 50mm in height and 50mm across; individuals are not readily distinguished and are arranged rather irregularly around the common exhalent openings; colony is smooth and transparent with red individuals showing through.

**Habitat** on stones, rocks, seaweeds, sponges and shells, etc., from the lower shore down to about 50m.

**Distribution** Mediterranean, Atlantic, English Channel and North Sea. N.B. it is possible that this species may prove to be a variant of *Aplidium nordmanni* (not illustrated), which differs from it principally because it forms flat-topped, squat, encrusting colonies without club-shaped growths.

magnified
calcareous
spicule

*Didemnum*
*maculosum*

### *Didemnum maculosum* (Milne Edwards)

Sessile; individuals growing in rough, leathery colonies which form irregular, up to 1mm high, encrusting growths about 2mm thick and up to 40mm across; individuals not easy to distinguish; 5–8 individuals share a common exhalent opening.

**Colour** purple-yellow-white colonies contain calcareous spicules with up to 15 rays and are sometimes marked with purple lines.

**Habitat** under stones and on seaweeds, especially laminarian holdfasts (see pages 32–33), from lower shore down to deeper water.

**Distribution** Mediterranean, Atlantic, English Channel and North Sea, but not further north than England.

### Ciona intestinalis (Linnaeus)

Sessile; solitary individuals with soft, cylindrical, retractile bodies reaching up to 120mm in height; inhalent opening terminal, with exhalent opening close by; branchial region longer than gastric region.

**Colour** transparent with yellow-green hues, edges of openings are yellow.

**Habitat** often growing in great numbers on rocks, piers and piles as well as on seaweeds from lower shore down to 500m.

**Distribution** Mediterranean, Atlantic, English Channel, North Sea and west Baltic.

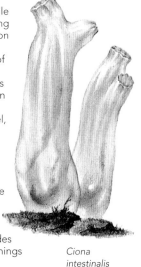

### Diazona violacea Savigny

Sessile; individuals growing in globular or more flattened colonies reaching 200mm in height and 400mm across; individuals themselves may reach 50mm high, and the branchial region of about 20mm length usually protrudes from the colony; inhalent and exhalent openings are terminal and close together.

**Colour** translucent yellow-green.

**Habitat** attached to rocks and stones, often in strong currents, from 30–200m.

**Distribution** Mediterranean, Atlantic and English Channel.

*Ciona
intestinalis*

individual removed
from colony

*Diazona
violacea*

### Ascidiella aspersa (O. F. Müller)

Sessile; solitary, rough-looking individuals usually more than 60mm high, but may occasionally reach 130mm; inhalent opening terminal, exhalent opening about one-third of body height away.

**Colour** brown-grey-black.

**Habitat** in clay (where it may develop a stalk) or attached to stones and seaweeds or piles from lower shore down to 50m.

**Distribution** Mediterranean, Atlantic north to Shetland, English Charm, and west Baltic. N.B. sea-quirts of this and other species may grow on *aspersa*.

*Ascidiella
aspersa*

273

### Ascidia mentula O. F. Müller

Sessile; solitary individuals up to 100mm high, set in a thick, cartilage-like tunic; surface with swellings of low profile; exhalent opening more than half body length away from terminal, inhalent opening; branchial and gastric regions of similar size.

**Colour** translucent with green hue.

**Habitat** attached to rocks and sometimes shells from lower shore down to about 200m.

**Distribution** Mediterranean, Atlantic, English Channel, North Sea and west Baltic.

*Ascidia mentula*

### Phallusia mammillata (Cuvier)

Sessile; solitary individuals up to 140mm high, set in a thick, cartilage-like tunic with many conspicuous smooth swellings; inhalent opening terminal, exhalent opening less than half of the body length away from it.

**Colour** varies with depth from white to brown.

**Habitat** usually attached to stones buried in mud or clay from the lower shore down to about 180m.

**Distribution** Mediterranean, and Atlantic to English Channel.

*Phallusia mammillata*

### Styela clava Herdman

Sessile; solitary individuals up to 70mm high, set in a leathery, crinkled tunic; exhalent opening about one-quarter of body length from terminal inhalent opening.

**Colour** white-brown.

**Habitat** attached to stones, sometimes growing very close together; may be encrusted with seaweeds.

**Distribution** Mediterranean and Atlantic.

*Styela clava*

### Dendrodoa grossularia (van Beneden)

Sessile, solitary or aggregated; individuals may be flattened to squat domes, or fairly upright and cylindrical, reaching 25mm high and 12mm across; exhalent opening about one-third of body length away from terminal inhalent opening.

**Colour** red-brown.

**Habitat** on rocks or shells from lower shore down to 100m or more; in sheltered places it often aggregates in large numbers and is more cylindrical in shape.

**Distribution** Atlantic, English Channel, North Sea and west Baltic. N.B. avoid confusion with *Distomus variolosus* (opposite).

*Dendrodoa grossularia*

### *Distomus variolosus* Gaertner

Sessile and colonial; cylindrical-globular individuals reach up to 10mm high which are easily recognized within the colony, being clustered closely together; they aggregate by budding; inhalent opening of individual is terminal, exhalent opening is close by; tunic has rough surface.

**Colour** red-brown.

**Habitat** encrusting rocks and holdfasts of *Laminaria* (see pages 32–33), and hydroids on lower shore and in shallow water.

**Distribution** Atlantic and English Channel N.B. avoid confusion with *Dendrodoa grossularia* (see opposite).

*Distomus variolosus*

### *Botryllus schlosseri* (Pallas)
**Star Ascidian**

Sessile; individuals arranged in characteristic, star-like groups of 3–12 or more within the colony; colonies of various shapes and sizes, often flat, but sometimes thicker and fleshy; individuals about 2mm long and arranged around a common exhalent opening.

**Colour** very variable; often brown-yellow-green or reddish with 'stars' standing out in contrasting colours.

**Habitat** encrusting stones, rocks and seaweeds, and sometimes hydroids and other ascidians, on lower shore and in shallow water.

**Distribution** Mediterranean, Atlantic, English Channel and North Sea.

*Botryllus schlosseri*

### *Botrylloides leachi* (Savigny)

General appearance similar to *Botrylus schlosseri* (above) but individuals, which are about 1.5mm long, are arranged in irregular rows on either side of an elongated exhalent cavity; do not resemble stars.

**Colour** orange-yellow-grey.

**Habitat** encrusting stones and other ascidians on the lower shore and in shallow water.

**Distribution** Mediterranean, Atlantic, English Channel and North Sea.

*Botrylloides leachi*

### *Molgula tubulifera* Oersted

Sessile and solitary; rounded, soft body up to 30mm high, inhalent and exhalent openings both terminal; tunic covered with small fibrils and sometimes with sand grains.

**Colour** blue-green.

**Habitat** on a variety of substrates and in areas of normal and reduced salinity from the lower shore down to about 100m.

**Distribution** Atlantic, English Channel, North Sea and west Baltic. This species is is now regarded as different from the apparently similar *M. manhattensis* (not illustrated) of American north Atlantic waters.

*Molgula tubulifera*

275

In addition to possessing chordate features, vertebrates have a backbone composed of many articulated body units, or vertebrae, arranged segmentally along with the main body musculature. The distinct head has associated sense organs, and the brain is protected by a brain case of skeletal material. Vertebrates often have paired front (pectoral) and rear (pelvic) appendages.

In terms of known species (about 46,700 species), the vertebrates are but a fraction of the whole animal kingdom. However, they include many of the most familiar organisms, and they range from the lowly lampreys to Man himself. In this book there is room to treat only one section of the vertebrates, the fishes, although seabirds, and mammals such as the porpoise and seal, are also familiar marine organisms; Several classes of vertebrates may be described as fishes. These are the Agnatha (fishes lacking true jaws), which include the lampreys and hagfishes; the Placodermi (an extinct class of primitive jawed fishes); the Chondrichthyes (cartilaginous fishes lacking hard bones), which include the sharks and rays; and the Osteichthyes which includes all the higher bony fishes of which the salmon and the goldfish are examples. The following account will include all these groups apart from the Placodermi.

Although the agnathans show a low level of vertebrate development (simple, unpaired fins; poorly developed sense organs, etc.), they show special modifications to their parasitic way of life. These include the mouth and surrounding teeth, and the arrangement of the gills which may permit respiration even when part of the head is buried in the host fish's tissues.

The class Chondrichthyes is represented in European waters by a range of forms including the smaller dogfishes, various skates and rays, and the very large basking sharks. The sharks, unlike the agnathans mentioned above, are all rapid and powerful swimmers, often leading lives as voracious carnivores. They frequently prey on shoaling fishes like herring and mackerel. Some of the smaller species are scavengers, while the basking sharks are filter feeders, straining the surface waters for small, planktonic organisms in the manner of many whales. The skates and rays are bottom dwellers, feeding on invertebrates living on or in the sand and mud, as well as preying on flatfishes. The diagram opposite shows the main external features of a male shark. It will be noted that

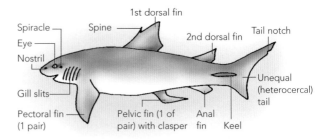

▶ External features of a generalized male shark

1st dorsal fin

Spiracle — Spine — 2nd dorsal fin — Tail notch
Eye —
Nostril —
Gill slits —
Pectoral fin — (1 pair)
Pelvic fin (1 of pair) with clasper
Anal fin
Keel
Unequal (heterocercal) tail

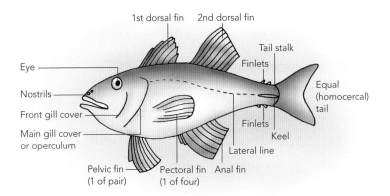

1st dorsal fin    2nd dorsal fin

Tail stalk

Eye

Finlets

Nostrils

Equal
(homocercal)
tail

Front gill cover

Main gill cover
or operculum

Finlets

Keel

Lateral line

Pelvic fin
(1 of pair)

Pectoral fin
(1 of four)

Anal fin

the pelvic fins are modified for use as claspers – structures which allow the transfer of sperms to the female, also true in skates and rays. Females lack claspers. The eggs are normally retained within the female body and, after fertilization, the embryos may develop there until they are sufficiently mature to lead an independent life after birth. In the case of certain species, such as the dogfishes, the embryos are extruded from the female body in special egg cases known as mermaids' purses (see page 315). These egg cases are attached to seaweeds, and eventually the juveniles hatch from them.

Most of the important food fishes in the world belong to the class Osteichthyes. The Osteichthyes are divided into a number of subgroups of which two concern us here. The first of these is the Chondrostei, which are rather primitive, with unequally developed tail fins. An example is the sturgeon, but there are many fossil forms. The second subgroup is the Teleostei, which includes the dominant fishes of the present day. They are distinguished by the equal tail, and by the swim bladder or buoyancy mechanism which enables them to maintain a particular position in the water without swimming. The diagram above indicates the principal external features of a teleost. Most female teleosts lay thousands of eggs which are fertilized externally by the male. In a number of cases the eggs float, and after fertilization they develop in the plankton as embryos, the juveniles feeding on the yolk supply of the egg. When this is exhausted they can often take small items of planktonic food. As adults they occupy almost all niches in the marine environment. Readers requiring more information should turn to Hardy, A. C. 1970; Muus, B. J. and Dalstrom, P. 1974; Lythgoe, J and Lythgoe, G. 1971 or Wheeler, A. 1969.

Because of variability both in life and after death, references to colour for identification purposes have been kept to a minimum.

▲ External
features of a
generalized
teleost

## Class Agnatha
# Lampreys and hagfishes
Vertebrates lacking true jaws

### *Petromyzon marinus* Linnaeus
**Sea Lamprey**
**Length** up to 900mm when adult.
**Head** bears small eyes; 7 pairs of gills; mouth opens to form an oval, sucker-like structure armed with many teeth arranged in several circlets; large centre tooth with 2 points.
**Body** long, slender and eel-like; 2 dorsal fins and 1 tail fin; skin smooth and lacking scales.
**Habitat** often near river mouths and fresh water, from shallow water down to 400m or more.

**Distribution** Mediterranean, Atlantic, English Channel, North Sea and Baltic.

*Petromyzon marinus*

### *Myxine glutinosa* Linnaeus
**Hagfish**
**Length** up to 400mm when adult.
**Head** bears no eyes; 2 barbels each side of terminal nostril and 2 more each side of mouth; 1 gill opening on each side.
**Body** eel-like; 1 fin continuous around posterior.
**Habitat** on muddy substrates in burrows from 25m downward; often attached to other fishes on which it preys.

**Distribution** Atlantic, English Channel and North Sea.

*Myxine glutinosa*

## Class Chondrichthyes
# Sharks, skates and rays

Vertebrates with true jaws and cartilaginous skeletons, and lacking bony fin rays. The mouth is on the under surface. The skin generally feels hard and rough when stroked from the tail towards the head. The tail usually has two unequal lobes.

### *Isurus oxyrinchus* Rafinesque
**Mako**
**Length** up to 4m when adult.
**Head** mouth has powerful jaws armed with large, sharply pointed teeth lacking cusps or notches; 5 large gill slits in front of pectoral fins.

*Isurus oxyrinchus*

**Body** typically shark-like; fairly slender; 2 dorsal fins, the first being immediately behind the posterior edge of the pectoral fin; tail has side-keels and a notch.

**Habitat** open water near the surface.

**Distribution** Atlantic north to the approaches to the English Channel.

### *Lamna nasus* (Bonnaterre)
**Porbeagle** or **Mackerel Shark**
**Length** up to 3.5m when adult.
**Head** teeth have notches or cusps
on either side of the main
triangular point.
**Body** similar to *Isurus
oxyrinchus* (opposite) but
with a fuller shape and upper
part of tail fin longer; dorsal fin begins
where pectoral fin bones terminate.
**Habitat** from shallow water down to about 150m.
**Distribution** Mediterranean, Atlantic, English Channel and
North Sea. N.B. illustration shows a female, identifiable by the
lack of claspers on the pelvic fins (see page 276).

*Lamna
nasus*

### *Cetorhinus maximus* (Gunnerus)
**Basking Shark**
**Length** up to 15m when adult; great size gives easy
identification.
**Head**     pointed
snout; tiny teeth;
small eyes; 5 long
gill slits in front of
the pectoral fin; gill
slits run from near the
top of the body almost
to the mid-line below;
gills have rakers which
strain plankton from the sea water passing through
them, so providing food.
**Body** 1st dorsal fin lies between pectoral and pelvic fins.
**Habitat** usually in open surface water,
migrating to deeper water in Winter.
**Distribution** Mediterranean, Atlantic,
English Channel, northern North Sea and
western Baltic.

*Cetorhinus
maximus*

### *Alopias vulpinus* (Bonnaterre)
**Thresher**
**Length** up to 4m when adult.
**Head** not as pointed as in some species.
**Body** the exceptionally long upper lobe of the tail fin gives
easy identification.
**Habitat** open surface water.
**Distribution** Mediterranean, Atlantic and
northern North Sea.

*C. maximus* profile

*Alopias
vulpinus*

279

### Scyliorhinus canicula (Linnaeus)
**Lesser-spotted Dogfish, Rough Hound** or **Rock Salmon**
**Length** up to 700mm when adult.
**Head** blunt, rounded snout; nostrils connected to the mouth by a conspicuous, fairly straight external groove visible on the underside of the snout; 5 pairs of small gill slits, the posterior 2 pairs overlap the pectoral fins; 1 pair of spiracles.
**Body** 2 dorsal fins, the second lying just behind the anal fin.
**Habitat** near or on sandy and muddy substrates, from shallow water down to 100m or more.
**Distribution** Mediterranean, Atlantic, English Channel and North Sea.

*Scyliorhinus canicula*

underside of snout

### Scyliorhinus stellaris (Linnaeus)
**Large-spotted Dogfish** or **Nurse Hound**
**Length** up to 1m when adult.
**Head** very similar to *S. canicula* (above); may be distinguished by the nostril grooves on the underside of the snout which do not connect the nostrils to the mouth, but terminate short of it and then run towards the mid-line almost meeting to form a w-shape.
**Body** 2nd dorsal fin begins over the centre of the anal fin.
**Habitat** often on rocky ground in sheltered places, from shallow water down to about 50m.
**Distribution** Mediterranean, Atlantic, English Channel and North Sea.

*Scyliorhinus stellaris*

underside of snout

### Mustelus mustelus (Linnaeus)
**Smooth Hound**
**Length** up to 1.5m when adult.
**Head** sharp, tapering snout; jaws with flat, unpointed teeth somewhat resembling diamond-shaped tiles and adapted for crushing the prey rather than tearing it to pieces; 5 gill slits, the last overlapping the pectoral fin.
**Body** 1st dorsal fin starts posterior to pectoral fins; 2nd dorsal fin starts just anterior to anal fin; conspicuous tail notch.
**Habitat** on soft substrates from 5–100m.
**Distribution** Mediterranean, Atlantic, English Channel and North Sea.

*Mustelus mustelus*

### Mustelus asterias Cloquet
**Stellate Smooth Hound**
(Not illustrated)
**Length** up to 2m when adult.
**Head** pointed.

Body not quite as slender as *M. mustelus* (opposite), but resembling it in most respects apart from the presence of white markings on the sides and back.
Habitat on sandy and muddy substrates, from shallow water down to about 150m.
Distribution Mediterranean, Atlantic, English Channel and North Sea.

### *Prionace glauca* (Linnaeus)
**Blue Shark**
Length up to 4m when adult.
Head pointed snout; triangular teeth with serrated edges; 5 pairs of small gill slits, the last pair overlapping the long blade-like pectoral fins.
Body 1st dorsal fin larger than 2nd; no keels; tail fin with notch on dorsal lobe; relatively smooth skin because scales are reduced in size.
Habitat open surface water.
Distribution Mediterranean, Atlantic north to Norway and the extreme west of the English Channel.

*Prionace glauca*

### *Galeorhinus galeus* (Linnaeus)
**Tope**
Length up to 2m when adult.
Head conspicuously pointed snout; sharp, pointed teeth with small accessory points on the posterior sides; 5 gill slits, the last pair overlapping the large pectoral fins.
Body 1st dorsal fin lies between pectoral and pelvic fins and is considerably larger than the 2nd dorsal fin; 2nd dorsal fin lies slightly in front of the anal fin.
Habitat on gravel or sandy substrates, from shallow water down to about 250m.
Distribution Mediterranean, Atlantic, English Channel and North Sea.

*Galeorhinus galeus*

### *Squalus acanthias* (Linnaeus)
**Spiny Dogfish**
Length up to 1.2m when adult.
Head rounded snout; spiracle behind eyes; 5 gill slits, all in front of pectoral fins.
Body 1st and 2nd dorsal fins with an anterior spine; pelvic fins lie in front of 2nd dorsal fin; anal fins lacking; no keels or notch in tail.
Habitat on a variety of substrates, from shallow water downward.
Distribution Mediterranean, Atlantic, English Channel and North Sea.

*Squalus acanthias*

**Squatina squatina** (Linnaeus)
**Monkfish** or **Angel Shark**
Length up to 2m when adult.
   Head blunt in outline with 2 nostrils and small barbels; conspicuous spiracles lie behind eyes.
      Body flattened and resembling that of a ray; very large pectoral fins, pelvic fins smaller; 2 dorsal fins; no anal fin; upper lobe of tail is smaller than lower.
   Habitat on sand and gravel, sometimes partly buried, usually in shallow water and down to 100m.
   Distribution Mediterranean, Atlantic and English Channel.
N.B. two other species of *Squatina* occur in the Mediterranean and nearby Atlantic but are not illustrated. *S. aculeata* (Linnaeus) is distinguished from *S. squatina* by a row of conspicuous spines running down its back; *S. oculata* Bonaparte has large black spots on the pectoral fins and on the body towards the tail.

*Squatina squatina*

**Torpedo marmorata** Risso
**Electric Ray**
Length up to 600mm when adult.
Head bears conspicuous spiracles behind the eyes, fringed on their inner margins by small protrusions of skin.
Body banjo-shaped; pectoral fins form edges of body; smaller pelvic fins; anal fins lacking; smooth skin lacks scales.
Habitat on sandy substrates, sometimes partly buried, from shallow water down to 200m.
   Distribution Mediterranean, Atlantic north to south-west Britain and Ireland and English Channel. N.B. this fish can give quite powerful shocks when touched, by discharging its electric organs which are modified muscles. Two similar species occur in the same regions as well as further afield. *T. torpedo* (Linnaeus) may be distinguished by its possession of about 5 large, blue spots set in dark rings on the predominantly brown back. The spiracles are generally large and unfringed. *T. nobiliana* Bonaparte is dark grey to black all over and has unfringed spiracles.

*Torpedo marmorata*

**Raja clavata** Linnaeus
**Thornback Ray**
Length up to 800mm when adult.
Head bears conspicuous spiracles.
Body pectoral fins form edges of the body which are wing-like; pelvic fins set close-by; dorsal fins lacking; very rough skin with conspicuous spines on the upper surface of the body and tail.
Habitat on soft substrates from shallow water down to 100m or more.
Distribution Mediterranean, Atlantic, English Channel, North Sea and west Baltic.

### *Raja batis* (Linnaeus)
**Common Skate**
**Length** up to 2m when adult.
**Head** snout tends to be more pointed than that of *R. clavata* (opposite).
**Body** row of conspicuous spines along the tail.
**Habitat** sandy and muddy substrates, from shallow water down to about 500m.
**Distribution** rarely in Mediterranean, Atlantic, English Channel and North Sea.
N.B. about ten other species of Raja occur in the European area and may be difficult to identify. Wheeler, A. 1969 gives identification details.

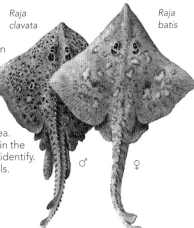

*Raja clavata*

*Raja batis*

♂    ♀

### *Dasyatis pastinaca* (Linnaeus)
**Common Stingray**
**Length** up to 2.3m overall.
**Head** fairly pointed snout; eyes smaller than spiracle situated just behind them.
**Body** pectoral fins form edge of body; pelvic fins adjacent; no dorsal fins; long tapering tail about 1.5 times length of body and bearing conspicuous toothed spine; female illustrated here.
**Habitat** on soft substrates sometimes partly buried and often in sheltered places in shallow water down to 50m.
**Distribution** Atlantic, English Channel, southern North Sea and rarely west Baltic. N.B. the tail spine is associated with a venom sac and extremely painful (but not fatal) wounds maybe inflicted.

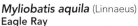

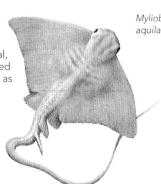

*Dasyatis pastinaca*

### *Myliobatis aquila* (Linnaeus)
**Eagle Ray**
**Length** up to 2m overall.
**Head** more distinct from body than in *Dasyatis pastinaca* (above); eyes and spiracles set sideways.
**Body** diamond-shaped, with pectoral, pelvic and small dorsal fins, the last situated just in front of tail spine; tail about twice as long as body.
**Habitat** in surface water and on soft substrates down to about 250m.
**Distribution** Mediterranean, Atlantic north to Scotland and English Channel.

*Myliobatis aquila*

# Class Osteichthyes Bony fishes

Fishes with true jaws and bony skeletons

## Subclass Actinopterygii

Infraclass Chondrostei

Primitive fishes with bony rays in the fins, the gill openings covered by a gill cover or operculum, and with unequally developed tail lobes.

### *Acipenser sturio* Linnaeus
**Sturgeon**
**Length** up to 1.5m when adult, though occasionally up to 4m.
**Head** pointed snout with 4 barbels; oval sucker-mouth.
**Body** has 5 rows of large bony plates running along it; upper lobe of tail larger than lower.
**Habitat** on sandy and muddy substrates, often in brackish water, and penetrating far up rivers to spawn.
**Distribution** Mediterranean (especially Adriatic), Atlantic, English Channel, North Sea and Baltic. N.B. a closely related species *A. huso* Linnaeus (not illustrated) may be found in the eastern Mediterranean and Adriatic. Adults reach up to 6m in length and have crescent-shaped mouths and much smaller bony plates.

*Acipenser sturio*

## Infraclass Teleostei

More advanced fishes having bony rays in the fins, gill openings covered by an operculum, equally developed tails and an internal swim bladder.

### *Sprattus sprattus* (Linnaeus)
**Sprat**
**Length** up to 150mm when adult.
**Head** operculum with radial ridges.
**Body** 1 dorsal fin set slightly back from mid-line; pelvic fins begin just in front of dorsal fin; forked tail fin; sharp scales give belly a serrated keel; large scales easily lost from body.
**Habitat** open water in shoals coming in shore towards Winter; from 10–150m.

*Sprattus sprattus*

**Distribution** Mediterranean, Atlantic, English Channel, North Sea and Baltic. N.B. the Mediterranean and Baltic races do not generally interbreed with those found in the Atlantic, English Channel and North Sea.

***Clupea harengus*** Linnaeus
**Herring**

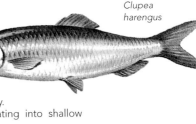
*Clupea harengus*

**Length** up to 400mm when adult.
**Head** operculum smooth.
**Body** very similar to *Sprattus sprattus* (opposite) but with dorsal fin in middle of back; pelvic fins begin just behind start of dorsal fin; sharp scales on belly form a weak ridge only.
**Habitat** open water in vast shoals migrating into shallow inshore waters towards Winter.
**Distribution** Atlantic north from Biscay, English Channel, North Sea and Baltic. N.B. several races of herring are recognized from the north Atlantic and Baltic. Mixed shoals of young Herrings and young Sprats are known as whitebait. Much scientific work has been carried out on the habits of this important food fish, see Muus, B. J. and Dahlstrom, P. 1974 or Hardy, A. C. 1970 for brief reviews.

***Sardina pilchardus*** (Walbaum)
**Sardine** (when small) or **Pilchard** (when large)

*Sardina pilchardus*

**Length** up to 260mm when adult.
**Head** operculum with radial ridges.
**Body** 1 dorsal fin set a little forward of the mid-line; pelvic fins start below the middle of the dorsal fin; last rays of the anal fin are long; forked tail; no sharp keel; large scales; may show a few darker spots on flanks.
**Habitat** in coastal waters in Spring and Summer, moving to deeper water in Winter.
**Distribution** Atlantic north to Ireland and northern England, English Channel and North Sea.

***Salmo salar*** Linnaeus
**Salmon**

*Salmo salar* old

**Length** male up to 1.5m, female up to 1.2m when adult.
**Head** large jaws, the upper extending back to the posterior margin of the eye; in breeding condition the jaws of old males may become hooked. ♂
**Body** 2 dorsal fins, 2nd lacks bony rays and is called the *adipose* fin.
**Habitat** open sea, migrating via coastal waters to spawn in rivers.
**Distribution** Atlantic north from Biscay, English Channel, North Sea and Baltic. N.B. migration and breeding habits are frequently described in books.

*Salmo Salar*

♀

***Salmo trutta*** Linnaeus
**Sea Trout**
(Not illustrated)
**Length** up to 1m when adult.
**Head** upper jaw usually reaches back beyond posterior limit of eye.
**Body** similar to *S. salar* (illustrated on page 285), but deeper, and tail stalk deeper.
**Habitat** coastal waters, but entering rivers to spawn.
**Distribution** Atlantic north of Spain, English Channel, North Sea and Baltic.

Note on the freshwater fishes Apart from several species which occur in both fresh and sea water throughout the European area and which will be treated below, a number of typical freshwater fishes occur in the Baltic where the salinity is low enough for them to exist. They include *Abramis brama* (Linnaeus), Bream; *Alburnus alburnus* (Linnaeus), Bleak; *Leuciscus idbarus* (Linnaeus), Ide; *Rutilus rutilus* (Linnaeus), Roach; *Tinca tinca* (Linnaeus), Tench; *Perca fluviatilis* (Linnaeus), Perch; and *Stizostedion lucioperca* (Linnaeus), Pike-perch. A description of these fishes is beyond the scope of this book and reference should be made to an appropriate text such as *The Collins Guide to Freshwater Fishes* by Muus, B. and Dahlstrom, P. 1971.

*Esox lucius*

***Esox lucius*** Linnaeus
**Pike**
**Length** up to 1m when adult.
**Head** the extended snout and long jaws give easy identification.
**Body** single dorsal fin set far back on the powerful body; slightly forked tail.
**Habitat** freshwater lakes and rivers and sea areas of low salinity.
**Distribution** throughout the European area and Baltic.

*Anguilla anguilla*
older form

***Anguilla anguilla***
(Linnaeus)
**Common Eel**
**Length** up to 1.4m when adult.
**Head** upper jaw shorter than lower one; gill slit.
**Body** dorsal fin starts about a quarter of the way back from snout and well behind pectoral fins; no pelvic fins; minute scales embedded in skin.

*Anguilla anguilla*
young form

**Habitat** in lakes, rivers and coastal waters as well as in the ocean depths.
**Distribution** Atlantic, English Channel, North Sea and Baltic.

N.B. the lengthy migration made by this species to the Sargasso Sea to spawn is described in many books dealing with fishes. The illustrations show two forms, the younger or yellow form which has a swollen head when viewed from above and the older or silver form which has lost its swollen headed appearance and is ready for its journey to the Sargasso Sea. There are many intermediate forms.

### *Conger conger* (Linnaeus)
**Conger Eel**
**Length** up to 2m
when adult.
**Head** large
mouth; upper
jaw fractionally longer
than lower jaw; 1 gill slit.

*Conger conger*

**Body** more massive and powerful than *Anguilla anguilla* (opposite); dorsal fin starts just behind pectoral fins; no pelvic fins; no scales.
**Habitat** among rocks and wrecks in shallow and very deep water.
**Distribution** Mediterranean, Atlantic, English Channel, North Sea and west Baltic. N.B. this is a truly marine species.

### *Muraena helena* Linnaeus
**Moray Eel**
**Length** up to 1.3m when adult.
**Head** long, powerful jaws; gill opening surrounded by a black ring.
**Body** no pectoral or pelvic fins; dorsal fin starts just in front of gill opening.

*Muraena helena*

**Habitat** in cracks of rocks and reefs and in old amphorae (as illustrated) in both shallow and deeper water.
**Distribution** Mediterranean and Atlantic north to Biscay.

### *Belone belone* (Linnaeus)
**Garfish** or **Garpike**
**Length** up to 800mm when adult.
**Head** fine, tapering jaws, the upper being the shorter.
**Body** long, slender; 1 dorsal fin set far back near forked tail; pectoral and pelvic fins; anal fin situated under dorsal fin; no finlets between these fins and tail.
**Habitat** open surface water in shoals, occasionally inshore.
**Distribution** Mediterranean, Atlantic, English
Channel, North Sea
and Baltic.

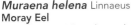

### *Scomberesox saurus* (Walbaum)
**Skipper**
(Not illustrated)

*Belone belone*

**Length** up to 500mm when adult. Similar to *Belone belone* (above) in appearance and details; about 5 finlets between dorsal fin and tail and about 6 finlets between anal fin and tail (see page 277).

### *Trisopterus luscus* (Linnaeus)
**Pout, Whiting** or **Bib**

*Trisopterus luscus*

**Length** up to 300mm when adult.
**Head** upper jaws slightly longer than lower; conspicuous barbel on lower jaw.
**Body** deep; 3 dorsal fins; pelvic fins lie in front of pectoral fins which have blackish spot at base; anus set vertically beneath centre of 1st dorsal fin; 2 anal fins; very short spaces between dorsal fins and anal fins.
**Habitat** on rocky and sandy substrates from shallow water down to about 100m.
**Distribution** Atlantic north from Biscay, English Channel, North Sea and west Baltic.

### *Trisopterus minutus* (Linnaeus)
**Poor Cod**

*Trisopterus minutus*

**Length** up to 200mm when adult.
**Head** upper jaw longer than lower jaw.
**Body** similar to *T. luscus* (above) but not quite as deep and lacking vertical colour bands; anus lies slightly further back behind the hind part of the 1st dorsal fin.
**Habitat** around piles, piers and rocks from shallow water down to 100m.
**Distribution** Atlantic north from Brittany, English Channel, North Sea and west Baltic.

### *Pollachius pollachius* (Linnaeus)
**Pollack**

*Pollachius pollachius*

**Length** up to 1.2m when adult.
**Head** upper jaw shorter than lower jaw, which lacks barbel.
**Body** 3 dorsal fins and 2 anal fins with distinct spaces between each; lateral line (see fig. 49) curves over pectoral fins and meets horizontal plane of fish just below start of 2nd dorsal fin; 1st anal fin starts below 1st dorsal fin and ends short of terminal part of 2nd dorsal fin.
**Habitat** among piers and rocks inshore in shallow water and down to 100m.
**Distribution** Atlantic, west Mediterranean, English Channel, northern North Sea and west Baltic.

### *Pollachius virens* (Linnaeus)
**Saithe** or **Coley**

*Pollachius virens*

**Length** usually up to 600mm, but may be twice this size.
**Head** minute barbel on lower jaw (may not be visible).
**Body** similar to *P. pollachius* (above) but distinguished by pale, nearly straight lateral line; 1st anal fin starts vertically beneath the start of 2nd dorsal fin.

Habitat among piers and rocks inshore and down to 200m.
Distribution Atlantic north from Biscay, English Channel, North Sea and west Baltic.

### Gadus morhua Linnaeus
**Cod**
**Length** up to 1m when adult.
**Head** lower jaw shorter than upper; conspicuous barbel.
**Body** lateral line curves over pectoral fin and does not straighten out until it passes under the trailing edge of 2nd dorsal fin; 1st anal fin starts more or less directly under start of 2nd dorsal (this is similar to the Saithe, but curved lateral line and barbel identify Cod).
**Habitat** on soft substrates from shallow water down to 600m.
**Distribution** Atlantic north from Biscay, English Channel, North Sea and Baltic.

*Gadus morhua*

### Molva molva
(Linnaeus)
**Ling**
**Length** up to 1m when adult.
**Head** lower jaw shorter than upper and bearing simple barbel.
**Body** long, gradually tapering to tail.
**Habitat** among rocks from 10–400m.
**Distribution** Atlantic from Biscay northwards.

*Molva molva*

### Gaidropsarus mediterraneus (Linnaeus)
**Shore Rockling**
**Length** up to 250mm when adult.
**Head** 3 barbels; upper jaw does not reach back beyond eye.
**Body** 1st dorsal fin has pronounced leading ray, others reduced; 2nd dorsal fin long; long anal fin.
**Habitat** among rocks, in pools and shallow water down to 30m.
**Distribution** Mediterranean, Atlantic and English Channel.

*Gaidropsarus mediterraneus*

### Ciliata mustela (Linnaeus)
**Five-bearded Rockling**
**Length** up to 200mm when adult.
**Head** 5 barbels.
**Body** similar to *Gaidropsarus mediterraneus* (above).
**Habitat** on the shore and in shallow water in sandy places.
**Distribution** Atlantic north to Portugal, English Channel, North Sea and west Baltic. N.B. several other species of cod-like fishes and Rocklings occur in the European area. These include the Whiting (see page 290); for descriptions of Haddock, Hake and Three- and Four-bearded Rocklings, etc., see Wheeler, A. 1969.

*Ciliata mustela*

### *Merlangius merlangus* (Linnaeus)
**Whiting**

*Merlangius merlangus*

**Length** up to 500mm when adult.
**Head** upper jaw longer than lower jaw, which normally lacks barbel.
**Body** similar in shape to both *Trisopterus minutus* and *T. luscus* on page 288, but less deep; 1st anal fin starts vertically beneath centre of 1st dorsal fin and terminates level with end of 2nd dorsal fin.
**Colour** blue-green-brown above, sides silvery with grey-black spot at the base of pectoral fin.
**Habitat** near soft substrates from 20–150m.
**Distribution** western Mediterranean, Atlantic north from Spain, English Channel, North Sea and west Baltic.

### *Hippocampus guttulatus* Leach
**Seahorse**
**Height** up to 150mm when adult.
**Head** snout is more than one-third of the total head length.
**Body** 'mane' or crest of appendages runs from behind the eyes to the dorsal fin on the 'back'; the horse-like appearance gives easy identification.
**Habitat** among seaweeds and sea-grasses especially *Posidonia* (see page 65).
**Distribution** Mediterranean and Atlantic north to English Channel.

### *Hippocampus hippocampus* Linnaeus
**Seahorse**
**Height** up to 150mm.
**Head** snout is one-third or less of the total head length.
**Body** similar to *H. guttulatus* (above).
**Habitat** as for *H. guttulatus*.
**Distribution** Mediterranean and Atlantic north to English Channel.

*Hippocampus guttulatus*

*Hippocampus hippocampus*

### *Syngnathus acus* Linnaeus
**Greater Pipefish**
**Length** up to 500mm when adult.

*Syngnathus acus*

**Head** tapering snout occupies more than half total head length; small hump over gill opening.
**Body** very long and slender; pectoral fins present; about 18 body rings between pectoral fin base and start of dorsal fin; small tail fin.
**Habitat** among sand, pebbles and rocks in shallow water.
**Distribution** Mediterranean, Atlantic, English Channel, North Sea and west Baltic.

### *Syngnathus rostellatus* Nilsson
**Lesser Pipefish**
(Not illustrated)
**Length** up to 170mm when adult.
**Head** snout occupies less than half head length.
**Body** similar to *S. acus*; about 15 body rings between pectoral fin base and start of dorsal fin; tail fin.
**Habitat** among seaweeds in shallow water.
**Distribution** Atlantic north of Spain, English Channel, North Sea and west Baltic.

### *Syngnathus typhle*
Linnaeus
**Broad-nosed Pipefish**
**Length** up to 350mm when adult.
**Head** snout not tapering and nearly as tall along its length as rest of head, also very flattened sideways.
**Body** about 19 body rings between pectoral fin base and start of dorsal fin; tail fin.
**Habitat** often in brackish water among seaweeds and in shallow places.
**Distribution** Mediterranean, Atlantic, English Channel, North Sea and Baltic.

*Syngnathus typhle*

### *Syngnathus phlegon*
Risso
**Length** up to 200mm when adult.
**Head** snout more than half total head length.
**Body** pectoral fin; dorsal surface very rough when stroked from back to front; tail fin.
**Colour** varies from bluish to yellow with darker spots or stripes.
**Habitat** open surface water and down to 600m.
**Distribution** Mediterranean and Atlantic.

*Entelurus
aequoreus*

*Syngnathus
phlegon*

### *Entelurus aequoreus* (Linnaeus)
**Snake Pipefish**
**Length** up to 600mm when adult.
**Head** snout more than half total head length.
**Body** very long and slender; no pectoral fin; very small tail fin; anus situated below and towards hind end of dorsal fin.
**Habitat** among seaweeds in shallow water down to about 25m.
**Distribution** Atlantic, English Channel, North Sea and west Baltic.

291

### *Nerophis ophidion* (Linnaeus)
**Straight-nosed Pipefish**
**Length** up to 300mm when adult.
**Head** snout about half total head length with a straight ridge up to the eyes.
**Body** no pectoral or tail fins; anus situated below and towards the front end of the dorsal fin.
**Colour** variable; dark to pale green sometimes with vertical lines or spots.
**Habitat** among seaweeds in very shallow water.
**Distribution** Mediterranean, Atlantic, English Channel, North Sea and west Baltic.

*Nerophis ophidion*

### *Zeus faber* Linnaeus
**John Dory**
**Length** up to 400mm when adult.
**Head** has extendable mouth.
**Body** dorsal fin with long anterior rays and edged on each side by spiny scales; anterior part of anal fin has conspicuous rays; this fin is also edged by spiny scales; conspicuous blackish spot on side of body.
**Habitat** among rocks and seaweeds from shallow water down to 200m.
**Distribution** Mediterranean, Atlantic, English Channel and rarely in the North Sea and west Baltic.

*Zeus faber*

### *Chelon labrosus* (Risso)
**Thick-lipped Grey Mullet**
**Length** up to 600mm when adult.
**Head** relatively small mouth at the front; very thick lip on upper jaw; small teeth.
**Body** 1st dorsal fin with 4 conspicuous rays; large scales.
**Habitat** among rocks and seaweeds in shallow water and estuaries, often near sewage outfalls.
**Distribution** Mediterranean, Atlantic, English Channel and southern North Sea. N.B. several other species of Grey Mullet

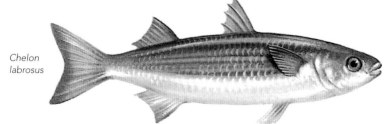

*Chelon labrosus*

occur in European waters, see Wheeler, A. 1969 or Riedl, R. 1963. The Grey Mullet is not a close relative of the Red Mullet *Mullus surmuletus* (see page 296).

### *Atherina presbyter* Cuvier
**Sand Smelt**
**Length** up to 150mm when adult.
**Head** has large mouth.
**Body** narrow, 2 clearly separated dorsal fins; forked tail; anal fin starts a little in front of 2nd dorsal fin; characteristic silvery line along sides.
**Habitat** in coastal waters, estuaries and in brackish water.
**Distribution** western Mediterranean, Atlantic, English Channel and North Sea.

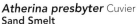

*Atherina presbyter*

### *Sphyraena sphyraena* (Linnaeus)
**Barracuda**
**Length** up to 500mm when adult.
**Head** has strong jaws, upper shorter than lower; many sharp teeth.
**Body** narrow, powerful, 2 dorsal fins clearly separated; anal fin below 2nd dorsal fin; adults usually have dark, vertical markings.
**Habitat** in open water often over sandy substrates; normally in shoals.
**Distribution** Mediterranean and Atlantic north to Biscay.

*Sphyraena sphyraena*

### *Dicentrarchus labrax* (Linnaeus)
**Bass**
**Length** up to 800mm when adult.
**Head** front gill cover has toothed hind edge; dark mark on main gill cover which has 2 spines.
**Body** long, ovoid; 2 separated dorsal fins each about the same length, the first with prominent rays.
**Habitat** over sand, shingle and rocks from shallow water down to about 100m, sometimes in estuaries.
**Distribution** Mediterranean, Atlantic, English Channel, North Sea and Baltic.

*Dicentrarchus labrax*

### *Polyprion americanum* Bloch & Schneider
**Wreckfish**
**Length** up to 2m when adult.
**Head** large; upper jaw shorter than lower; shape of forehead over eyes is important in identification: there is a slight dip in the outline just above the eyes; there are swellings between the eyes and on the brow; front gill cover is toothed on hind edge; main gill cover has a conspicuous horizontal ridge running across it approximately level with the eye.
**Body** dorsal fin bilobed with large spines at front.
**Habitat** among floating objects, wreckage, seaweeds, etc., from shallow water down to 200m and sometimes much deeper.
**Distribution** Mediterranean, Atlantic north to Ireland and English Channel.

*Polyprion
americanum*

### *Serranus cabrilla* (Linnaeus)
**Comber**
**Length** up to 250mm when adult.
**Head** front gill cover has toothed hind edge; main gill cover has 2 points on it near the horizontal mid-line of the fish.
**Body** elongated; one dorsal fin with prominent rays in the anterior part; about 8 conspicuous, upright brown bands and 2 blue-green horizontal lines as well as yellow-orange markings.
**Habitat** among rocks, sand and sea-grasses, from shallow water, down to 50m; sometimes much deeper.

*Serranus
cabrilla*

**Distribution** Mediterranean, Atlantic north to western English Channel.

### *Boops boops* (Linnaeus)
**Bogue**
**Length** up to 200mm when adult.
**Head** small mouth; teeth notched, upper teeth with 4 points and lower with 5 points each; front gill cover not toothed; no spines on the main gill cover.
**Body** shallow; upper and lower edges have similar curvature; dorsal fin long, anterior rays more spiny than posterior; anal

fin has spiny anterior rays and terminates approximately level with hind edge of dorsal fin; lateral line is a dark curved mark following the profile of the fish's back; about 3 darkish lines lie below it; blackish spot at pectoral fin base.

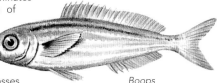

**Habitat** among rocks, sand and sea-grasses from shallow water down to 150m.

*Boops boops*

**Distribution** Mediterranean and Atlantic north to Biscay, and occasionally further north.

### *Pagellus bogaraveo* (Brünnich)
**Red Sea-bream**

**Length** up to 350mm when adult.
**Head** mouth on lower side; front teeth small and sharp; rear ones small but blunter; eye large and conspicuous; gill covers untoothed and lacking spines.

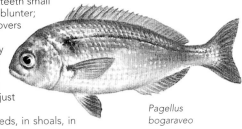

**Body** ovoid; upper edge of body more strongly convex than lower edge; in most adults there is a conspicuous blackish spot at the anterior end of the lateral line just behind the gill.

*Pagellus bogaraveo*

**Habitat** among rocks and seaweeds, in shoals, in shallow water and down to 250m or more.

**Distribution** Atlantic north to Ireland and occasionally as far as Scandinavia and northern North Sea.

### *Dentex dentex* (Linnaeus)
**Dentex**

**Length** up to 1m when adult.
**Head** large; mouth on extreme lower side; front teeth long and conspicuous, rear teeth smaller; eye small; gill covers untoothed and lacking spines.
**Body** ovoid; upper edge more strongly convex than lower edge.
**Colour** somewhat variable, but generally as illustrated with about 5 vertical darker bands which are usually indistinct; very large specimens greater than 1m may be reddish all over.
**Habitat** among rocks from shallow water down to 200m.
**Distribution** Mediterranean and Atlantic north to Biscay.

*Dentex dentex*

### *Sparus aurata* Linnaeus
**Gilthead**
**Length** up to 700mm when adult.
**Head** steep snout and 'forehead'; upper jaw protrudes slightly; large lips; strong sharp pointed teeth in front of jaw, with crushing teeth behind; gill covers untoothed and lacking spines.

**Body** ovoid; darkish patch where lateral line meets gill cover.

**Habitat** in shallow water among rocks; will tolerate brackish conditions.

**Distribution** Mediterranean and Atlantic north to Biscay.

*Sparus aurata*

### *Mullus surmuletus* Linnaeus
**Red Mullet**
**Length** up to 300mm when adult.
**Head** mouth low; 2 long sensory barbels on the lower jaw; 2 conspicuous scales run from end of jaw to below the eye.

*Mullus surmuletus*

**Body** long and somewhat compressed.

**Habitat** among rocks and sand from shallow water down to 100m.

**Distribution** Mediterranean and Atlantic; north to Scotland. N.B. this fish is not a close relation of the Grey Mullet. It can change its colour considerably according to situation, stress, time of day, etc. *Mullus barbatus* Linnaeus is distinguished by its more vertically inclined 'forehead' and 3 conspicuous scales which run from the end of the jaw to below the eye. Authorities are not all in agreement as to the status of this species, which some feel is a variety of *M. surmuletus*.

*M. barbatus*

### *Sciaena umbra* Linnaeus
**Corb** or **Brown Meagre**
**Length** up to 500mm when adult.

**Head** large, rounded.

**Body** deep, ovoid; 2 dorsal fins, 1st with conspicuous spines, 2nd just connected to it by a fine piece of skin; 2nd spine of anal fin is conspicuous; tail fin straight-ended in adults.

*Sciaena umbra*

**Habitat** among rocks and caves, seaweeds and sea-grasses often in shoals; generally in shallow water.

**Distribution** Mediterranean and Atlantic north to Biscay.

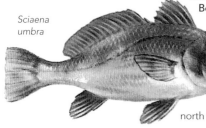

### *Trachurus trachurus* (Linnaeus)
**Scad** or **Horse Mackerel**
**Length** up to 350mm when adult.
**Head** large, pointed, with
a conspicuous mouth.
**Body** slim; 2 dorsal
fins, 1st taller than 2nd;
deeply cleft tail fin; anal fin
preceded by 2 spines; pectoral fin
wing-like and when folded, stretches

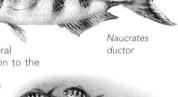

*Trachurus
trachurus*

back to the 2 spines; lateral line curves back from the gill and
straightens out under the 2nd dorsal fin; large, pointed scales
form a 'keel' along the lateral line on each side of the fish.
**Habitat** in shoals in open water from the surface down to
200m or more; sometimes close inshore over sand.
**Distribution** Mediterranean, Atlantic, English Channel, North
Sea and west Baltic. N.B. this fish is easily distinguished from
the true Mackerel by its lack of finlets (see page 277).

### *Naucrates ductor* (Linnaeus)
**Pilotfish**
**Length** up to 400mm when adult.
**Head** rounded.
**Body** shallow, ovoid; 1st dorsal fin
reduced to a few isolated spines in
front of 2nd dorsal fin; forked tail; anal fin
preceded by 2 spines; keels on tail stalk over lateral
line; dark, vertical bands on the body extend on to the

*Naucrates
ductor*

anal, dorsal and tail fins; tail edged with white.
**Habitat** open water, often accompanying large
fish such as sharks, as well as turtles, floating
objects, etc.; young specimens may keep
company with jellyfishes.
**Distribution** Mediterranean, Atlantic and
English Channel.

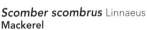

### *Scomber scombrus* Linnaeus
**Mackerel**
**Length** up to 400mm when adult.
**Head** pointed.
**Body** long, shallow, but plump; 2 dorsal fins, 1st has 11–13
spines; anal fin lies under 2nd dorsal fin;
about 5 finlets (see page 277) between
2nd dorsal fin and tail fin, and
between anal fin and tail
fin; deeply cleft tail;
stripes along back.
**Habitat** pelagic in
shoals, coming inshore
in Summer and Autumn.
**Distribution** Mediterranean, Atlantic,
English Channel, North Sea and west Baltic.

*Scomber
scombrus*

### *Scomber japonicus* Houttuym
**Spanish Mackerel**
**Length** up to 300mm when adult.
**Head** pointed.
**Body** very similar to *S. scombrus* (page 297); 9–10 spines in first dorsal fin; beneath lateral line there is a clear row of darker spots; golden-yellow stripe runs along side of fish.

**Habitat** pelagic in shoals.

**Distribution** Mediterranean and Atlantic north to English Channel.

*Scomber japonicus*

### *Sparisoma cretense* (Linnaeus)
**Parrotfish**
**Length** up to 300mm when adult.
**Head** blunt; small mouth reveals apparently massive teeth (composed of a number of fused teeth) which resemble a parrot's beak.
**Body** ovoid; two colour varieties are known, the one illustrated probably being the female; there is a grey form (probably the male) which has a blackish patch behind the gill cover.

*Sparisoma cretense*

**Habitat** among rocks and seaweeds, often in shallow water.
**Distribution** Mediterranean and Atlantic north to Portugal.

### *Balistes carolinensis* Gmelin
**Triggerfish**
**Length** up to 250mm when adult.
**Head** large; small mouth.
**Body** deep, ovoid-diamond shape; 1st dorsal fin has massive front spine followed closely by another, and then by a 3rd a little way back; 2nd dorsal fin; over anal fin, top and bottom rays of tail fin may be extended back as trailers.
**Colour** variable, from green-brown-blue.
**Habitat** among rocks and seaweeds in coastal waters.
**Distribution** Mediterranean, Atlantic north to Ireland, and English Channel. N.B. the 1st ray of the 1st dorsal fin may be raised and locked into position. It is often used to wedge the fish in crevices.

*Balistes carolinensis*

**Note on the wrasses** This large group of fishes includes many common and familiar inhabitants of coastal waters. Some of them are most brilliantly coloured and patterned. Often the appearance varies with the time of year, courtship, etc., and frequently the males and females are differently marked. The basic body shape is ovoid and somewhat elongate. The mouth is at the extreme tip of the head. The front gill cover is often toothed, but the main cover never bears spines. The dorsal fin is often divided into two sections with the front few rays of the first part spiny. In this group of fishes there are a number of hermaphrodite species that start life as females and then, as they age, change their sex to become males. This is a familiar condition in many invertebrates but is less common in vertebrates.

*L. mixtus* mature
in breeding
condition

**Labrus mixtus** Linnaeus
**Cuckoo Wrasse**

*Labrus mixtus*
mature

**Length** up to 350mm when adult.
**Head** front gill cover untoothed.
**Body** fairly long and shallow; ♂
front of dorsal fin sometimes
slightly lower than rear.
**Colour** immature males
orange-red colour; mature
males have bright blue heads and sides,
blue extending the front of dorsal fin and
blue stripe on tip of tail fin; white heads indicate
courtship coloration; females orange-red
with 3 dark patches on rear of back.
**Habitat** among rocks from about 10m
downward.
**Distribution** Mediterranean, Atlantic and English
Channel.

*L. mixtus*

♀

**Labrus bergylta** Ascanius **( =L. maculatus)**
**Ballan Wrasse**

*Labrus
bergylta*

**Length** up to 400mm when adult.
**Head** large; large lips; slightly humped
'forehead'; gill covers untoothed and
lacking spines.
**Body** front of dorsal fin may be slightly
lower than rear when fully erect; large
scales usually having a paler centre and darker
posterior edge.
**Colour** varies according to age and
stress; two colour forms of *L. bergylta*
are illustrated here.
**Habitat** among rocks and seaweeds
in coastal waters down to about 20m.
**Distribution** western Mediterranean,
Atlantic, English Channel, North Sea and west Baltic.

*Labrus
bergylta*

299

### *Coris julis* (Linnaeus)
**Rainbow Wrasse**
**Length** up to 200mm when adult.
**Head** pointed snout with mouth at tip; eyes are relatively inconspicuous.

♂

**Body** longer and less deep than many wrasses; 1st 2 rays of long dorsal fin spiny in male; front of dorsal fin about same height as the rear; like a number of wrasses this species is hermaphrodite; the first

*Coris julis*

(female) phase is distinguished by a blue spot on the bottom edge of the gill cover and a yellowish line running from the snout towards the tail; the second (male) is distinguished by the spiny rays and a blackish spot on the dorsal fin; the male also has a

♀

characteristic double saw-toothed orange pattern along the side and a dark mark about half way along the body.
**Habitat** among rocks and seaweeds from shallow water down to 120m.
**Distribution** Mediterranean and Atlantic north to Biscay.

### *Crenilabrus melops* (Linnaeus)
**Corkwing Wrasse**
**Length** up to 200mm when adult.
**Head** front gill cover toothed.

**Body** fuller shape than most wrasses; lateral line less curved and straightening out towards rear of dorsal fin, about 3 spiny rays at front of anal fin.
**Colour** very variable according to age and conditions (male illustrated with breeding coloration); a dark spot in front of the tail fin, over or just below the

♂

*Crenilabrus melops* breeding

lateral line, is a characteristic feature of this species.
**Habitat** among rocks and seaweeds in rock pools and in shallow water.
**Distribution** Mediterranean, Atlantic, English Channel, North Sea and west Baltic.

### *Centrolabrus exoletus* (Linnaeus)
**Rock Cook**
**Length** up to 150mm when adult.
**Head** front gill cover toothed.
**Body** about 5 spiny rays at front of anal fin.
**Colour** varies with age and conditions; most characteristic distinguishing feature is the dark bar on the tail and the paler

♂

*Centrolabrus exoletus* breeding

trailing edge; male illustrated with breeding coloration.
**Habitat** among rocks and seaweeds in shallow water.
**Distribution** Atlantic from Biscay northwards, English Channel and North Sea.

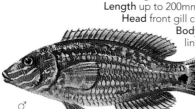

### *Centrolabrus rupestris* (Linnaeus)
**Goldsinny**
**Length** up to 150mm when adult.
**Head** fine serrations on the front gill cover.
**Body** back somewhat flattened; lateral line follows a more or less straight course under the front part of the dorsal fin but curves downwards towards the tail under the hind lobe of the dorsal fin; reddish-brown colour distinguishes this species; blackish spot in front of the tail fin, well above the lateral line; sometimes a similar spot on the front of the dorsal fin.
**Habitat** among rocks from shallow water down to about 30m.
**Distribution** Mediterranean, Atlantic and English Channel.

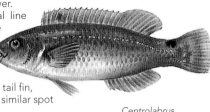

*Centrolabrus rupestris*

### *Trachinus draco* Linnaeus
**Greater Weever**
**Length** up to 350mm when adult.
**Head** flattened sideways; large oblique mouth; eyes almost on top of head; about 2 small spines on upper edge of eye socket; large, conspicuous poison spine on main gill cover.
**Body** notch in pectoral fin; front dorsal fin has about 3 spines and a blackish spot towards the front; 2nd dorsal fin long; diagonal stripes on side of body.
**Habitat** often buried in sand, from shallow water down to 100m.
**Distribution** Mediterranean, Atlantic, English Channel and North Sea.

*Trachinus draco*

### *Trachinus vipera* Cuvier
**Lesser Weaver**
**Length** up to 120mm when adult.
**Head** similar to *T. draco* (above), but lacking spines above eye socket; poison spine on gill cover.
**Body** also similar, though deeper; pectoral fin lacks notch; 1st dorsal fin is black with about 6 spines.
**Habitat** and **Distribution** similar to *T. draco*. N.B. the spines on the dorsal fins and on the gill covers of these two species are venomous and cause exceedingly painful wounds. Allow wounds to bleed freely before cleaning them carefully.

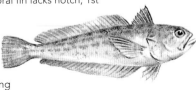

*Trachinus vipera*

301

### *Uranoscopus scaber* Linnaeus
**Star Gazer**
**Length** up to 250mm when adult.
**Head** large, with eyes on upper surface; small process on lower jaw.
**Body** similar in shape to the weevers; pectoral fin bases in front of the beginning of the dorsal fin; large sculptured gill covers with 1 sharp venomous spine behind; 1st dorsal fin has about 4 rays.
**Habitat** often buried in sand, in shallow water.
**Distribution** Mediterranean and Atlantic north to Spain.

*Uranoscopus scaber*

### *Callionymus lyra* Linnaeus
**Dragonet**
**Length** up to 250mm when adult.
**Head** longish pointed snout; mouth set low; quite large lips; frog-like eyes; front gill cover with 4 spines.
**Body** long, thin and somewhat compressed vertically; 1st dorsal fin very long and sail-like in the mature male, but no longer than the 2nd in the juvenile male and female.
**Habitat** on soft substrates from shallow water down to about 100m.
**Distribution** Mediterranean, Atlantic, English Channel, North Sea and west Baltic. N.B. spawning takes place in mid water with the female swimming vertically beside the male as he displays to her (see inset illustration). Several closely related species occur in the European area (see Wheeler, A. 1969).

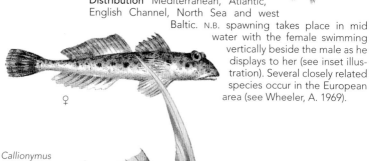

♀

*Callionymus lyra*   ♂

### *Hyperoplus lanceolatus* (Lesauvage)
### *( =Ammodytes lanceolatus)*

**Length** up to 320mm when adult.
**Head** the dark mark on the snout just in front of the eye gives easy identification; upper jaw is not extendable.
**Body** long and slender; the oblique folds in the skin lack scales (hand lens useful here).
**Habitat** on sandy substrates, from shallow water down to 150m.
**Distribution** Atlantic, English Channel, North Sea and west Baltic.

### *Ammodytes tobianus* Linnaeus *( =A. lancea)*
**Lesser Sandeel**

*Hyperoplus lanceolatus*

**Length** up to 200mm when adult.
**Head** similar to *Hyperoplus lanceolatus* (above) but lacking black spot on snout, in front of eye; upper jaw is extendable.
**Body** also similar to *H. lanceolatus* but there are scales present on the oblique folds in the skin (hand lens useful).
**Habitat** on sandy bottoms in shallow water down to about 30m.
**Distribution** Atlantic north from Portugal, English Channel, North Sea and west Baltic. N.B. several other species of sand-eel occur in the area. See Lythgoe, J. and Lythgoe, G. 1971, or Wheeler, A. 1969.

### *Pholis gunnellus* (Linnaeus)
**Butterfish**

*Ammodytes tobianus*

**Length** up to 200mm when adult.
**Head** small, rounded; thick lips.
**Body** compressed sideways; very long dorsal fin with about 11 dark spots ringed by pale pigment; dorsal fin about twice length of anal fin; pelvic fins level with pectoral fins and greatly reduced; distinct tail fin.
**Colour** predominantly brown, as illustrated, and grey.
**Habitat** in mud, sand and among rocks, from the lower shore down to about 50m.
**Distribution** Atlantic north from English Channel, English Channel, North Sea and west Baltic.

*Pholis gunnellus*

### *Zoarces viviparus* (Linnaeus)
**Eelpout** or **Viviparous Blenny**
**Length** up to 400mm when adult.
**Body** longish, tapered, and more massive than *Pholis gunnellus* (page 303); reduced pelvic fins in front of pectorals; long dorsal fin is continuous with anal fin, but there is a notch in the dorsal fin just before the tail; regular darker patches along the back.
**Colour** illustration shows a male with breeding coloration; otherwise pectoral fins are brownish and fringed with orange-yellow.
**Habitat** on sand, mud and under stones in shallow water and sometimes down to about 50m.
**Distribution** Atlantic north from mid-Wales and Northern Ireland, eastern English Channel, North Sea and Baltic.

*Zoarces viviparus*

**Note on the blennies** This large group of fishes has a number of European representatives. They lack scales, and their bodies are typically long and rounded in section. They bear superficial similarity to the gobies (see pages 306–307), but can be distinguished from them because: **1.** their pelvic fins, which are set well in front of the pectorals, consist of a few rays only and are not united to form a ventral sucker as in the gobies; **2.** gobies have scales; **3.** the dorsal fin is usually single but may be divided into 2 lobes by a notch or 'dip' whereas in the gobies the 1st dorsal fin is usually clearly separated from the 2nd dorsal fin.

### *Liophyrs pholis* (Linnaeus)
**Shanny**
**Length** up to 150mm when adult.
**Head** sharply rising snout with 'brow' over eye; no head tentacles.
**Body** typical dorsal fin with a poorly defined blotch of darker pigment towards the front.
**Colour** ground colours vary from brown to green, but the male becomes much darker in the breeding season.
**Habitat** among rocks and stones on the lower shore and in shallow water.
**Distribution** Atlantic north from Portugal, English Channel and North Sea.

*Liophyrs pholis*

### *Blennius ocellaris* Linnaeus
**Butterfly Blenny**
**Length** up to 170mm when adult.
**Head** bears 2 short divided appendages above the eye and 1 more very small one on either side of the front of the dorsal fin.
**Body** fairly deep; compressed sideways; 1st ray of dorsal fin very long and there is a conspicuous dark mark surrounded by a white ring towards the rear of the front lobe.
**Habitat** usually in shallow water, often associated with coralline algae, e.g. *Lithothamnion* (not illustrated).
**Distribution** Mediterranean, Atlantic and English Channel.

*Blennius ocellaris*

### *Parablennius gattorugine* (Linnaeus)
**Tompot Blenny**
**Length** up to 250mm when adult.
**Head** eyes well up towards 'brow'; 1 much-divided tentacle above each eye socket.
**Body** fairly deep; typical dorsal fin with spiny rays at front, rays in posterior part are the tallest.
**Habitat** among stones and kelps from lower shore downward.
**Distribution** Mediterranean, Atlantic and English Channel.

*Parablennius gattorugine*

### *Coryphoblennius galerita* (Linnaeus) *( =Blennius montagui)*
**Montagu's Blenny**
**Length** up to 80mm when adult.
**Head** a fringe of tentacles runs crossways over the head just behind the eyes and a crest of small projections (larger in the male) runs back towards the start of the dorsal fin.
**Body** slender, tapering; front and rear lobes of dorsal fin separated by a dip.
**Habitat** among rocks and seaweeds, usually in pools on the lower shore and in shallow water.
**Distribution** Mediterranean (rarely), Atlantic north to south-west Britain and western English Channel.

*Coryphoblennius galerita*

Note on the gobies These small, generally inshore fishes form a large group which includes about 50 European species. To distinguish them from the blennies, see the preceding page. For species not treated below, see Wheeler, A. 1969 and Lithgoe, J. and Lithgoe, G. 1971.

### *Gobius niger* Linnaeus *( =G. jozo)*
**Black Goby**
**Length** up to 150mm when adult.
**Body** slightly fuller than many gobies; 1st dorsal fin taller and more crest-like in male (illustrated) than in female; pectoral fins have a few short, spiny, upper rays which do not reach up to the base of the 1st dorsal fin; tail stalk relatively short.
**Colour** dark, and with various markings.
**Habitat** over soft substrates and among sea-grasses in shallow water and down to 60m; can tolerate brackish water.
**Distribution** Mediterranean, Atlantic, English Channel, North Sea and west Baltic.

*Gobius niger*

### *Gobius paganellus* Linnaeus
**Rock Goby**
**Length** up to 120mm when adult.
**Body** thickset; 1st dorsal fin characterized by a band of orange, or a paler brown stripe, and having leading 4 rays about the same height; pectoral fins have several conspicuous, free, spiny upper rays which may reach to the base of the 1st dorsal fin; tail stalk relatively short.
**Habitat** among seaweeds in rock pools on lower shore and in shallow water.
**Distribution** Mediterranean, Atlantic north to Ireland and Scotland and western English Channel.

*Gobius paganellus*

### *Thorogobius ephippiatus* (Lowe)
**Leopard-spot Goby**
**Length** up to 130mm when adult.
**Body** fairly slender; 1st dorsal fin rounded, front ray tallest; pectoral fins lack spiny upper rays; tail stalk quite long; characteristic blotches of red-brown pigment, becoming larger towards the rear; blackish spot at hind base of 1st dorsal fin.
**Habitat** among rocks, extreme lower shore and shallow water down to about 35m.
**Distribution** Mediterranean and Atlantic north to south-west Britain.

*Thorogobius ephippiatus*

### *Pomatoschistus minutus* (Pallas) *( =Gobius minutus)*
**Sand Goby**
**Length** up to 90mm when adult.
**Body** slender; 1st dorsal fin has its anterior rays tallest; small, dark blue spot at hind end of dorsal fin in male; upper edge of pectoral fins not spiny; tail stalk quite long; characteristic markings include about 4 indistinct vertical bands of darker pigment along the sides.
**Habitat** in shallow, sandy places.
**Distribution** Mediterranean, Atlantic, English Channel, North Sea and west Baltic.

*Pomatoschistus minutus*

### *Lesueurigobius friesii* (Collett) *( =Gobius friesii)*
**Fries' Goby**
**Length** up to 120mm when adult.
**Body** slender; this genus is immediately distinguished by the short tail stalk and diamond-shaped tail fin; anterior rays of 1st dorsal fin spiny and tall; no spiny rays on pectoral fin.
**Habitat** in deep water from 50m downward, where it may share the burrows of *Nephrops norvegicus* (Norway Lobster, see page 225); quite rare but may be taken in dredges, etc.
**Distribution** Mediterranean, Atlantic, English Channel and North Sea.

*Lesueurigobius friesii*

*L. friesii* and *Nephrops norvegicus*

### *Gobiusculus flavescens* (Fabricius)
**Two-spotted Goby**
**Length** about 60mm.
**Head** eyes quite widely separated.
**Body** 2nd dorsal fin slightly longer than 1st, which has about 7 rays; these fins are well separated; upper edges of pectoral fins not spiny; conspicuous spots below 1st dorsal fin and at base of tail; long tail stalk.
**Habitat** above the seabed, among seaweeds, etc., in shallow places.
**Distribution** Atlantic north from Spain, English Channel, North Sea and Baltic.

*Gobiusculus flavescens*

### *Eutrigla gurnardus* (Linnaeus)
### Grey Gurnard

**Length** up to 400mm when adult (rather smaller in Mediterranean).

**Head** sharply pointed snout; 3 or 4 spines above mouth.

**Body** tapers towards tail; 1st dorsal fin spiny and well separated from 2nd; anal fin starts a little behind 2nd dorsal fin and is shorter; pectoral fin rays divided; lowermost 3 rays hang down and are used as sense organs to test substrate; remainder form a more normally rounded fin which does not reach start of anal fin; short spine above pectoral fin; pelvic fins start about level with pectoral fins, sides of dorsal fins edged by row of spines, and spine scales overlie lateral line; Mediterranean specimens are more brown-red and have a dark mark on 1st dorsal fin.

**Habitat** on soft substrates and among rocky outcrops, from shallow water down to 200m.

**Distribution** Mediterranean, Atlantic, English Channel, North Sea and west Baltic

*Eutrigla
gurnardus*

### *Aspitrigla cuculus* (Linnaeus)
### Red Gurnard

**Length** up to 300mm, when adult.

**Head** pointed snout somewhat 'dished' in front of eye and with about 3 spines on either side.

**Body** tapers towards tail; 1st dorsal fin tall in front and sail-like, with spiny rays; anal fin in about same relative position and of same length as 2nd dorsal fin; pectoral fins divided, with 3 sensory rays hanging down; rounded swimming section (see above) reaches back as far as start of anal fin; several small spines above start of pectoral fin; pelvic fins below pectoral fins; sides of dorsal fins edged by row of spines; lateral line covered by vertically arranged, smooth, scaly ridges.

*Aspitrigla
cuculus*

**Habitat** generally over soft substrates, from shallow water down to about 250m.

**Distribution** Mediterranean, Atlantic, English Channel, North Sea and west Baltic.

### *Myoxocephalus scorpius* (Linnaeus)
### ( =*Cottus scorpius*)
**Father Lasher** or **Short-spined Sea-scorpion**
**Length** up to 200mm when adult.
**Head** large, spiny and somewhat
flattened; large mouth and jaws;
front gill cover has 2 spines and
its lower edge is extended as a
flap which reaches under chin.
**Body** rounded, tapering towards tail; 2
spiny dorsal fins; anal fin starts iust behind start of
2nd dorsal fin and ends just before it; pectoral fin extends back
to start of anal fin; pelvic fins small with about 3 rays.
**Colour** varies according to conditions and season; here a
male is illustrated in breeding coloration; the upper surface of
the female is usually grey-brown with darker markings, and
the under surface is orange.
**Habitat** usually over soft substrates and among seaweeds, in
shallow water and down to 60m.
**Distribution** Atlantic north from Biscay, English Channel,
North Sea and west Baltic.

*Myoxocephalus
scorpius
breeding*

### *Scorpaena porcus* Linnaeus
**Scorpionfish**
**Length** up to 250mm when adult.
**Head** large and spiny; large mouth; feather-
like tentacles over eye; no tentacles
under chin; gill covers spiny.
**Body** fairly full; long dorsal fin, front
lobe having spiny rays, rear part with
smooth outline, and slightly more
lobed; short anal fin terminates about
level with end of dorsal fin; camouflage of
this fish makes it difficult to see against substrates.
**Habitat** among rocks in shallow water.
**Distribution** Mediterranean and Atlantic north to Biscay. N.B.
the spines on the dorsal fin and gill covers are very poisonous
and can inject a powerful venom if touched. The affected part
of the body should be treated with very hot water to relieve
pain and then well cleaned.

*Scorpaena
porcus*

### *Agonus cataphractus* (Linnaeus)
**Pogge** or **Armed Bullhead**
**Length** up to 150mm when adult.
**Head** large with barbels beneath.
**Body** tapers to a very slender tail
stalk; large pectoral fin; 2
dorsal fins; body entirely
cased in bony plates.
**Habitat** on soft substrates in
shallow water and down to about 500m;
often burrows under mud or sand; may occur in estuaries.
**Distribution** Atlantic north from south-west Britain. English
Channel, North Sea and west Baltic.

*Agonus
cataphractus*

309

### *Cyclopterus lumpus* Linnaeus
**Lump Sucker** or **Sea Hen**
**Length** up to 550mm when adult.
**Head** mouth and eyes relatively small.
**Body** large, thickset and traversed by 4 rows of bony plates; no scales; 2 dorsal fins, but in older specimens the 1st appears reduced and incorporated in the body; anal fin about equal in length to 2nd dorsal fin; pelvic fins form a sucker under the body.

*Cyclopterus lumpus mature ♂*

**Colour** adult male in breeding coloration illustrated, otherwise males and females are grey-blue below; small immature specimens may be greenish.

**Habitat** usually on rocky substrates from extreme lower shore down to about 300m; may be attached to rocks by the ventral sucker formed from the pelvic fins.

**Distribution** Atlantic north from Portugal, English Channel and the North Sea.

### *Liparis montagui* (Donovan)
**Montagu's Sea-snail**
**Length** up to 70mm when adult.
**Head** rounded, blunt.
**Body** relatively long and tapering towards tail; hind parts somewhat compressed sideways; body covered with slippery, loose skin; long dorsal fin; shorter anal fin, not connected to tail fin; pectoral fins somewhat extended under chin; pelvic fins form a ventral sucker.

*Liparis montagui*

**Habitat** among rocks and stones, from the middle shore down to about 30m.
**Distribution** Atlantic north from south-west Britain, English Channel, North Sea and west Baltic. N.B. a very similar fish *Liparis liparis* (Linnaeus) (not illustrated), Common Sea-snail, differs from *L. montagui* principally in that its anal fin joins the tail fin.

### *Gasterosteus aculeatus* Linnaeus
**Three-spined Stickle-back**
**Length** up to 60mm when adult.
**Head** pointed with small mouth and large eyes.
**Body** long and laterally compressed; 3 dorsal spines precede dorsal fin which is slightly larger than the anal fin; pelvic fins reduced to 1 spine and 1 ray.
**Colour** male in breeding coloration illustrated, otherwise the sides are silver and the underparts whitish.

*Gasterosteus aculeatus breeding*

**Habitat** in shallow water, often among seaweeds where the nest is built.

**Distribution** throughout Europe in conditions of varying salinity, Atlantic north from Biscay, English Channel, North Sea and Baltic. N.B. this familiar freshwater fish also inhabits brackish and fully marine areas.

### *Spinachia spinachia* (Linnaeus)
**Fifteen-spined Stickle-back**
**Length** up to 160mm when adult.
**Head** tapering snout with tiny mouth.
**Body** long and tapering; pelvic fins reduced to 1 spine and 1 small ray; anal fin about level with, and equal in length to, the dorsal fin, which is preceded by about 15 spines; extremely fine tail stalk.
**Habitat** this sea fish is often found among rocks in inshore waters and estuaries, in pools on the lower shore, and down to about 10m.
**Distribution** Atlantic north from Biscay, English Channel, North Sea and Baltic.

*Spinachia spinachia*

### *Lepadogaster lepadogaster* (Bonnaterre)
**Clingfish** or **Cornish Sucker**
**Length** up to 70mm when adult.
**Head** long pointed snout; quite long jaws; upper jaw bigger than lower which is covered when the mouth is shut; tentacle over nostril.
**Body** flat below and curved above; anal fin shorter than dorsal fin and, like it, linked to the tail fin; rounded pectoral fins; pelvic fin modified to form a sucker.
**Habitat** often in pools among rocks and seaweeds from the middle shore down to a few metres.
**Distribution** Mediterranean, Atlantic north to Scotland and western English Channel.

*Lepadogaster lepadogaster*

### *Lophius piscatorius* (Linnaeus)
**Anglerfish**
**Length** up to 2m when adult.
**Head** huge, grotesque, with an enormous crescent-shaped mouth just on the upper surface; very much broader than rest of body; about 3 long spines borne in a row from above the upper lip to between the eyes, the foremost spine having a fleshy tip which acts as a lure to bring small fishes within reach of the mouth; gill slits directly behind pectoral fins.
**Body** somewhat flattened top to bottom and tapering towards the tail; dorsal fin well back.
**Habitat** from shallow water down to about 500m.
**Distribution** Mediterranean, Atlantic, English Channel and North Sea.

*Lophius piscatorius*

311

N.B. all the fishes on this plate are illustrated with their dorsal fins uppermost.

### *Scophthalmus maximus* (Linnaeus)
**Turbot**
**Length** up to 600mm when adult.
**Head** long dorsal fin commences between mouth and eyes.
**Body** rounded; left-hand side uppermost; pectoral and pelvic fins; separate anal fin; scales lacking; upper side bears a number of hard raised platelets which give a rough surface; lateral line curves steeply over the pectoral fin.

*Scophthalmus maximus*

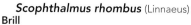

**Colour** upper side varies according to substrate and the need for camouflage; tail fin spotted; underside pale.

**Habitat** on mud, sand and gravel substrates, from shallow water down to about 100m.

**Distribution** Mediterranean, Atlantic, English Channel, North Sea and Baltic.

### *Scophthalmus rhombus* (Linnaeus)
**Brill**
(Head end only illustrated) Very similar to *S. maximus* (above).
**Length** up to 500mm when adult.
**Body** lacks bony platelets on upper surface, but has scales; anterior rays of dorsal fin tufted.
**Colour** may be darker with fins lighter and the tail fin less spotted.
**Habitat** and **Distribution** similar to *S. maximus* but not penetrating as far into the Baltic.

*Scophthalmus rhombus*

### *Zeugopterus punctatus* (Bloch)
**Topknot**

*Zeugopterus punctatus*

**Length** up to 200mm when adult.
**Head** dorsal fin starts between mouth and eyes.
**Body** oval; left-hand side uppermost; pelvic fin attached to anal fin; belts of darker pigment extend either side of eye sockets and there is a dark patch posterior to the curve in the lateral line; the hind edges of the scales are toothed (hand lens useful here).

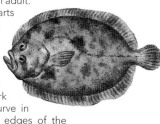

**Habitat** among rocks from shallow water down to about 40m.
**Distribution** Atlantic north from Biscay, English Channel, North Sea and west Baltic.

### *Platichthys flesus* (Linnaeus)
**Flounder**
**Length** up to 200mm when adult.
**Head** long dorsal fin commences near the eye.
**Body** less rounded than the Turbot, and with the right-hand side uppermost; pelvic and anal fins separate; row of hard knobs follows the outline of the body where it joins the dorsal and anal fins; shallow curve to the lateral line over pectoral fin; scales lack toothed edges.
**Colour** upper side green-brown, and may be speckled with darker spots or even orange; underside pale.
**Habitat** on soft substrates, from shallow water down to 50m; this species can tolerate brackish or fresh water, so it may be found far up-river also.
**Distribution** Mediterranean, Atlantic, English Channel, North Sea and Baltic.

*Platichthys flesus*

### *Pleuronectes platessa* Linnaeus
**Plaice**
**Length** up to 550mm when adult.
**Body** very similar in general outline to *Platichthys flesus* (above) but the tail is rounded rather than almost square cut; no hard bony knobs at edge of body under dorsal and anal fin bases; upper left side patterned with conspicuous orange blotches.
**Habitat** on soft substrates, from shallow water down to about 350m.
**Distribution** western Mediterranean, Atlantic, English Channel, North Sea and west Baltic.

*Pleuronectes platessa*

### *Solea solea* (Linnaeus)
**Sole**
**Length** up to 300mm when adult.
**Head** blunt, with mouth not at the front extremity.
**Body** elongated and with the right-hand side uppermost; long dorsal fin; pelvic fin just separate from anal fin; dorsal and anal fins reaching tail fin.
**Habitat** on soft substrates from shallow water, but also down to about 200m; sometimes in river mouths.
**Distribution** Mediterranean, Atlantic, English Channel, North Sea and west Baltic.

*Solea solea*

A number of other species of flatfish may be encountered in the European area, for which reference should be made to Lythgoe, J. and Lythgoe, G. 1971, Wheeler, A. 1969, or Riedl, R. 1963.

# FLOTSAM AND JETSAM

A variety of objects may be found cast up by the sea and they usually accumulate at the high tide mark to form the strand-line. In many cases they include the remains of dead or dying animals; sometimes wreckage with attached animals will be found. It must be remembered that on shingle beaches the objects that occur may have been pounded by the action of pebbles so altering their form. The following list is by no means exclusive.

egg mass of
*Buccinum undatum*

## Egg mass of
### *Buccinum undatum* Linnaeus

Spongy looking masses usually reaching about the size of an apple. These consist of a great number of egg capsules laid in coils which form a spherical mass. In most cases the individual egg capsules will be empty when found, for the embryos of this species develop inside the egg case after fertilization and hatch out when they are ready to lead an independent existence. Because of this, some people might reasonably argue that the capsules should be termed embryo capsules.

## Egg capsules of
### *Nucella lapillus* (Linnaeus)

Small, flask-like capsules about 7mm tall attached to rocks or shells. These are normally laid in groups and bear a very superficial resemblance to some small sponges. A number of other gastropods lay their eggs intertidally in this way.

egg capsules of
*Nucella lapillus*

## Egg mass of
### *Archidoris pseudoargus* (Rapp)

Coiled, gelatinous streamer-like egg mass. Close examination reveals that the individual egg capsules are embedded in a transparent jelly. The mass may reach several centimetres in length. A number of other nudibranchs lay egg masses similar to this.

## Egg capsule of
### *Eulalia viridis* (O. F. Müller)

(see page 123) Shaped like a miniature, greenish pear, this jelly-like object has an attachment strand at one end. It is usually about 25mm long. It is often found lying on sand on the middle and lower shore.

egg mass of
*Archidoris pseudoargus*

### Mermaids' Purses Egg capsules of
**Scyliorhinus canicula** (Linnaeus)
Capsule length about 60mm. These horny egg capsules are usually attached to seaweeds by means of the twisted tendrils which arise from each corner. Those that are washed up on the shore are generally empty because the embryo has hatched.

### Mermaids' Purses Egg capsules of
**Scyliorhinus stellaris** (Linnaeus)
(Not illustrated) Similar to the above, only about twice as long.

### Mermaids' Purses Egg capsules of
**Raja clavata** (Linnaeus)
(Not illustrated) Capsule length about 60mm. Similar to that of *Scyliorhinus canicula* above, but with four tapering points rather than tendrils and with a slightly hairy covering.

### Mermaids' Purses Egg capsules of
**Raja batis** Linnaeus
(Not illustrated) Capsule length about 130mm, with four points rather than tendrils. They are often covered with quite long, sticky hairs.

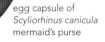

egg capsule of
*Scyliorhinus canicula*
mermaid's purse

### Hornwrack Fronds of the ectoproct
**Flustra foliacea** (Linnaeus)
Leafy fronds reaching up to 60mm long, often washed up after storms.

*Flustra
foliacea*

### Cuttlebone of
**Sepia officinalis** (Linnaeus)
(see also page 196) Length up to about 180mm. This structure forms the internal skeleton for the cuttlefish. It is a porous structure which in life can be partly filled with gas to form a buoyancy organ which regulates the vertical position of the cuttlefish in the water. After death and decay of the cuttlefish the bone is often washed up.

### Sea-balls
Matted, fibrous balls 4–60mm in diameter (see page 65).

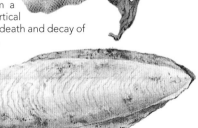

cuttle bone of
*Sepia officinalis*

315

# BIBLIOGRAPHY

The following list of reference works, which is by no means exclusive, includes a variety of monographs, fauna lists and guides to specific areas as well as a few scientific papers published in journals. The author has referred to these books in the course of preparing the present volume. In order to guide the reader further, * denotes references which are recommended for assisting in identification of particular groups of organisms; † denotes those which are fauna lists for particular areas; ‡ denotes those which are other important books of the field guide nature; and § denotes references of historical interest.

Admiralty Tide Tables *Vol 1 European Waters*. Hydrographic Department, Admiralty, Ministry of Defence, Taunton, Somerset. Contains tidal predictions for many major ports in Europe, and is of great use for planning field excursions. Published annually.

*Allen, J. A. 1967 *The Fauna of the Clyde Sea Area: Euphauslacea and Decapoda with an Illustrated key to the British species.* Scottish Marine Biological Association, Milport.

*Ball, I. R. and Reynoldson, 1981 *British Planarians. Synopses of the British Fauna (New Series) No. 19.* Published for the Linnean Society of London and the Estuarine and Coastal Sciences Association by the Field Studies Council, Preston Montford, Shrewsbury.

Ballantine, W. J. 1961. *A Biologically Defined Exposure Scale for the Comparative Descriptions of Rocky Shores. Field Studies, 1.,* 3, pages 1–19

*Barber, A. D. 2006 *Centipedes. Synopses of the British Fauna (New Series) No. 57.* Published for the Linnean Society of London and the Estuarine and Coastal Sciences Association by the Field Studies Council, Preston Montford, Shrewsbury.

*Berrill, N. J. 1950 *The Tunicata, with an account of the British Species.* Ray Society, London. A detailed guide to the ascidians.

Brayfield, A. E 1978. *Life in Sandy Shores. Studies in Biology 89.* Arnold, London. A short account on this subject.

†Bruce, J. R., Colman, J. S. and Jones, N. S. 1963 *Marine Fauna of the Isle of Man.* L.M.B.C. Memoir No. 36. Liverpool University Press.

*Brunberg, L. 1964 On the nemertean fauna of Danish Waters *Ophelia* 1, 77–112. A useful, illustrated guide to this group.

Campbell, A. C. and Dawes, J. 2005 *Encyclopedia of Underwater Life* Oxford University Press.

*Chambers, S. J. and Muir, R. I. 1997 *Polychaetes: British Chrysopetaloidea, Pisionoidea and Aphroditoidea. Synopsis of the British Fauna (New Series) No. 54.* Published for the Linnean Society of London and the Estuarine and Coastal Sciences Association by the Field Studies Council, Preston Montford, Shrewsbury.

*Clarke, A. M. and Dewney, M. E. 1992 *Starfishes of the Atlantic.* Chapman and Hall, London

*Cornelius, P. F. S. 1995 *North-West European Thecate Hydroids and their Medusae Pts 1 and 2. Synopses of the British Fauna (New Series) No. 50.* Published for the Linnean Society of London and the Estuarine and Coastal Sciences Association by the Field Studies Council, Preston Montford, Shrewsbury.

Cremona, J. 1988 *A Field Atlas of the Seashore* Cambridge University Press.

†Crothers, J. H. 1966 *Dale Fort Marine Fauna* (second edition). Field Studies Council, London. A fauna list for the Dale Peninsula and adjacent areas of Pembrokeshire, Wales.

*Crothers, J. H. and Crothers, M. 1983 *A Key to the Crabs and Crab-like Animals of British Inshore Waters* Field Studies Council, Preston Montford, Shrewsbury.

*Cuenot, L. 1922 *Faune de France*, 4; *Sipunculiens, Echiuriens, Priapuliens*. Paul Lechevalier, Paris. Provides a comprehensive reference text for the minor marine worms. Text in French.

*Darwin, C. 1851–1854 *A Monograph of the sub-class Cirripedia*. 2 vols. Ray Society, London. A classical reference work on barnacles.

‡Dickinson, C. I, 1963 *British Seaweeds*. The Kew Series, Eyre & Spottiswoode, London. A useful illustrated guide to marine algae.

Dixon, P. S. and Irvine, L. M. 1977. *Seaweeds of the Bntish Isles, Vol. 1 Rhodophyta, Part 1. Introduction, Nemaliales, Gigartinalies.* British Museum (Natural History), London

*Dobson, F. S. 1979 *Common British Lichens.* The Jarrold Nature Series, Jarrold, Norwich.

*Dobson, F. S. 1979 *Lichens. An Illustrated Guide.* The Richmond Publishing Co. Ltd.

*Dobson, F. S. 1997 *Lichens of the Rocky Shores.* The Richmond Publishing Co. Ltd.

Eltringham, S. K. 1971 *Life in Mud and Sand*. English Universities Press Ltd, London. A very useful account of the ecology of sandy and muddy habitats which have hitherto been somewhat neglected.

Fauchald, K. 1977 *The Polychaete Worms. Definitions and Keys to the Orders, Families and Genera.* Natural History Museum of Los Angeles County, Science Series, 28.

*Fauvel, P. 1923 *Faune de France 5: Polychètes errantes*. Paul Lechevalier, Paris. An essential text for the serious student of polychaetes. Together with the companion volume *Polychètes sedentaires* (listed below) it has superseded the Ray Society monograph by McIntosh. Text in French.

*Fauvel, P. 1927 *Faune de France 16: Polichètes sedentaires*. Paul Lechevalier, Paris.

*Ferry, B. W. and Sheard, J. W. 1969 Zonation of Supralittoral Lichens on Rocky Shores around the Dale Peninsula, Pembrokeshire (with key for their identification). *Field Studies, 3*, 1. 41–67

§ Forbes, E. 1841 *A History of British Starfishes*. J. Van Voorst, London. Includes an early discussion of zonation which is now only of historical interest.

Forsman, B. 1972 Evertebrater vid svenska osterjokusten *Zoologisk Revy*, 34, 32–56. An account of the bottom-dwelling invertebrates of the Baltic. Text in Swedish.

Fretter, V. and Graham, A. 1962 *British Prosobranch Molluscs*. Ray Society, London. An important work dealing with the biology of prosobranchs.

*George, J. G. and Hartman-Schröder, G. 1985 *Polychaetes: British Amphinomida, Spintherida and Eunicida. Synopses of the British Fauna (New Series) No. 32*. Published for the Linnean Society of London and the Estuarine and Coastal Sciences Association by the Field Studies Council, Preston Montford, Shrewsbury

*Gibbs. P. E. 1977. *British Sipunculans. Synopses of the British Fauna (New Series) No. 7*. Published for the Linnaean Society of London and the Estuarine and Coastal Sciences Association by the Field Studies Council, Preston Montford, Shrewsbury.

*Gibson, R. 1982 *British Nemerteans. Synopses of the British Fauna (New Series) No. 24*. Published for the Linnaean Society of London and the Estuarine and Coastal Sciences Association by the Field Studies Council, Preston Montford, Shrewsbury.

Gibson, R. 1972 *Nemerteans*. Hutchinson University Library, London. An excellent general account of the biology of nemertines.

§Gosse, P. H. 1860 *Actinologla Britannica. A History of the British Sea-anemones and Corals*. J. Van Voorst, London. The classical work on British sea-anemones and corals. To a great extent it has been superseded by the two Ray Society volumes by T. A. Stephenson.

*Graham, A. 1988 *Molluscs: Prosobranch and Pyramidellid Gastropods. Synopses of the British Fauna (New Series) No. 2*. Published for the Linnaean Society of London and the Estuarine and Coastal Sciences Association by the Field Studies Council, Preston Montford, Shrewsbury.

*Grimpe, G. and Wagler, E. (editors) 1927–1940 *Die Tierwelt der Nord-und Ostsee*. Akadmische Verlagsgesellschaft, Leipzig. An important series giving accounts of the fauna of the North Sea and the Baltic, containing many detailed identification keys and general ecological descriptions. Issued in many parts with many authors. Text in German.

Hardy, A. C. 1970 *The Open Sea: Part 2. Fish and Fisheries* (new edition). New Naturalist series, Collins, London. An important book on fish and bottom-dwelling animals.

Hardy, A. C 1971 *The Open Sea: Part 1. The World of Plankton* (revised edition). New Naturalist series, Collins, London. An excellent account of planktonic life.

Hawkins, S. J. and Jones, H. D. 1992 *Marine Field Course Guide 1 Rocky Shores* Marine Conservation Society, Immel Publications, London.

*Hayward, P. J. 1985 *Ctenostome Bryozoans. Synopses of the British Fauna (New Series) No.34* Published for the Linnaean Society of London and the Estuarine and Coastal Sciences Association by the Field Studies Council, Preston Montford, Shrewsbury.

*Hayward, P. J. and Ryland, J. S. 1985 *Cyclostome Bryozoans. Synopses of the British Fauna (New Series) No. 34* Published for the Linnaean Society of London and the Estuarine and Coastal Sciences Association by the Field Studies Council, Preston Montford, Shrewsbury.

*Hayward, P. J. and Ryland, J. S. 1998 *Cheilostomatous Bryozoa, Part 1. Aeteoidea-Cribrilinoidea. Synopses of the British Fauna (New Series) No. 10.* Published for the Linnaean Society of London and the Estuarine and Coastal Sciences Association by the Field Studies Council, Preston Montford, Shrewsbury.

*Hayward, P. J. and Ryland, J. S. editors 1998 *Handbook of the Marine Fauna of North-West Europe* Oxford University Press.

*Hayward, P. J. and Ryland, J. S. 1999 *Cheilostomatous Bryozoa, Part 2. Hippothooidea-Celleporoidea. Synopses of the British Fauna (New Series) No. 14.* Published for the Linnaean Society of London and the Estuarine and Coastal Sciences Association by the Field Studies Council, Preston Montford, Shrewsbury.

Heller, J. 1975 The taxonomy of some British *Littorina* species, with notes on their reproduction (Mollusca: Prosobrancia). *Zoological Journal of the Linnean Society, 56*, 2, 131–151.

*Hincks, T. 1868 *A History of British Hydroid Zoophytes*. 2 vols. J. Van Voorst, London. While much of the nomenclature has been changed since it was published, this book still remains an important work on hydroids for the serious student.

*Hiscock, S. 1979 *A Field Guide to the British Brown Seaweeds* Published for the Linnaean Society of London and the Estuarine and Coastal Sciences Association by the Field Studies Council, Preston Montford, Shrewsbury.

*Hiscock, S. 1986 *A Field Guide to the British Red Seaweeds* Published for the Linnaean Society of London and the Estuarine and Coastal Sciences Association by the Field Studies Council, Preston Montford, Shrewsbury.

*Hunnam, P. and Brown, G. 1975. Sublittoral Nudibranch Mollusca (Seaslugs) in Pembrokeshire Waters, *Field Studies 4*, 2, 131–159.

*Ingle, R. W. 1980 *British Crabs* British Museum (Natural History) and Oxford University Press.

*Ingle, R. W. 1996 *Shallow Water Crabs, Synopses of the British Fauna (New Series) No. 25.* Published for the Linnaean Society of

London and the Estuarine and Coastal Sciences Association by the Field Studies Council, Preston Montford, Shrewsbury.

*Ingle, R. W. and Christiansen, M. E. 2004 **Lobsters, Mud Shrimps and Anomuran Crabs, Synopses of the British Fauna (New Series) No. 55.** Published for the Linnaean Society of London and the Estuarine and Coastal Sciences Association by the Field Studies Council, Preston Montford, Shrewsbury.

*Jones, N. S. 1976 **British Cumaceans. Synopses of the British Fauna (New Series) No. 7.** Published for the Linnaean Society of London and the Estuarine and Coastal Sciences Association by the Field Studies Council, Preston Montford, Shrewsbury.

*King, P. E. 1974 **British Sea Spiders: Arthropoda: Pycnogonida. Synopses of the British Fauna (New Series) No. 5.** Published for the Linnaean Society of London and the Estuarine and Coastal Sciences Association by the Field Studies Council, Preston Montford, Shrewsbury.

*Kirkpatrick, P. A. and Pugh, P. R. 1984 **Siphonophores and Velellids. Synopses of the British Fauna (New Series) No 5.** Published for the Linnean Society of London and the Estuarine and Coastal Sciences Association by the Field Studies Council, Preston Montford, Shrewsbury.

†Laverack, M. S. and Blackler, M. 1974 **Fauna and Flora of St Andrews Bay.** Scottish Academic Press, Edinburgh and London.

Lewis, J. R. 1964 **The Ecology of Rocky Shores.** English Universities Press, London. See the Introduction for details.

*Lincoln, R. J. 1979 **British Marine Amphipoda: Gammaridea.** British Museum, London.

Little, C. 2000 **The Biology of Soft Shores and Estuaries.** Oxford University Press.

Little, C. and Kitching, J. A. 1996 **The Biology of Rocky Shores.** Oxford University Press.

*Lythgoe, J. and Lythgoe, G. 1971 **Fishes of the Sea.** Blandford, London. A comprehensive identification manual for fishes found in the coastal waters of northern Europe and the Mediterranean.

*Manuel, R. L. 1988 **British Anthozoa, Synopses of the British Fauna (New Series) No.18.** Published for the Linnaean Society of London and the Estuarine and Coastal Sciences Association by the Field Studies Council, Preston Montford, Shrewsbury.

†Marine Biological Association, 1957 **Plymouth Marine Fauna** (third edition). A fauna list for Plymouth and adjacent areas of southwest England, containing many important references.

*Matthews, G. 1953 A Key for use in the Identification of British Chitons. **Proceedings of the Malacological Society, London, 29**, 241–248.

*McIntosh, W. C. 1873–1923 A Monograph of the British Marine Annelids. 4 vols. Ray

Society, London. This extensive work, which is illustrated with many beautiful colour plates, is largely out of date as far as the nomenclature is concerned. It does, however, include at the start a comprehensive section on the nemerteans, which may be of use simply because there are few other alternatives. For normal use the reader is directed to Fauvel, P. in the **Faune de France** series.

*Millar. R. H. 1970 **British Ascidians; Tunicata: Ascidiacea. Synopses of the British Fauna (New Series) No. 1.** Published for the Linnaean Society of London and the Estuarine and Coastal Sciences Association by the Field Studies Council, Preston Montford, Shrewsbury.

*Moore, P. G. 1984 **The Fauna of the Clyde Sea Area. Crustacea: Amphipoda.** Occasional Publication 2, University Marine Station, Millport, Isle of Cumbrae

*Mortensen, T. 1927 **Handbook of the Echinoderms of the British Isles.** Oxford University Press. A very complete account of most of the north Atlantic species together with good identification keys.

*Muus, B. J. and Dahlstrom, P. 1974 **Collins Guide to the Sea Fishes of Britain and northwestern Europe.** Collins, London. This useful book provides identification details of many fishes and describes methods for their capture in addition to aspects of their biology.

*Naylor, E. 1972 **British Marine Isopods. Synopses of the British Fauna (New Series) No. 3.** Published for the Linnaeon Society of London and the Estuarine and Coastal Sciences Association by the Field Studies Council, Preston Montford, Shrewsbury.

†Newell, G. E. 1954 The Marine Fauna of Whitstable. **Annals and Magazine of Natural History Series 12**, 7, 321–350,

*Newell, G. E. and Newell, R. C. 1973 **Marine Plankton** (revised edition). Hutchinson, London. Provides a very useful and practical account of planktonic organisms together with many illustrations of species not treated in this guide book.

Newell, R. C. 1970 **Biology of Intertidal Animals.** Elek, London. Discusses in great detail many important aspects of life on the shore.

*Newton, L. 1931 **A Handbook of the British Seaweeds.** British Museum (Natural History) Publications, London. This book, although somewhat out of date, provides an important account of the marine algae, their distribution and structure, and can be used in conjunction with Parke and Dixon, 1976.

Nicol, J. A 1967 **The Biology of Marine Animals** (second edition). Pitman & Sons Ltd., London. Contains a great deal of information on the organization and physiology of the marine animals.

Nichols, D. 1969 **Echinoderm**s (fourth edition). Hutchinson University Library, London. A very good account of the structure and evolution of these animals.

*Organization for Economic Co-operation and Development (O.E.C.D) Catalogues of marine fouling organisms dealing with ascidians, barnacles, and Serpulids. H.M.S.O., London. These handy pamphlets contain useful identification keys and coloured illustrations.

†Parke, M. and Dixon, P. S. 1976 Checklist of British marine algae, third revision *Journal of the Marine Biological Association of the United Kingdom*, 56, 3, 527–594. Deals with further developments in the nomenclature of the algal flora.

*Picton, B. 1993 *A Field Guide to the Shallow-Water Echinoderms of the British Isles.* Immel Publications, London

*Prudhoe, S. 1982 *Polyclad Turbellarians. Synopses of the British Fauna (New Series) No. 26.* Published for the Linnaean Society of London and the Estuarine and Coastal Sciences Association by the Field Studies Council, Preston Montford, Shrewsbury.

Rainbow, P. S. 1984 *An Introduction to the Biology of British Littoral Barnacles* Field Studies, 6, 1–51

*Rasmussen. E. 1973 Systematics and ecology of the Isefjord marine fauna (Denmark) *Ophelia, 11,* 1–495. Includes descriptions and ecological data for many species occurring in the Kattegat region of Denmark

Reid, D. G. 1996 *Systematics and Evolution of Littorina.* Ray Society, London. 463pp.

*Russell, F. S. 1953 and 1970 *The Medusae of the British Isles.* 2 vols. Cambridge University Press. An up-to-date monograph on hydromedusae and scyphozoans with many illustrations, some in colour.

*Ryland, J. S. 1962 The biology and identification of intertidal Polyzoa. *Field studies, 1,* 4, 33–51.

Ryland, J S. 1970 *Bryozoans,* Hutchinson University Library, Lordon. An excellent account of the general biology of the ectoprocts.

*Ryland, J. S, 1974 A revised key of the identification of intertidal Bryozoa (Polyzoa) *Field Studies 4 1,* 77–86

*Smaldon, G., Holthius, L. B. and Fransen, C. H. J. M 1993 *Coastal Shrimps and Prawns. Synopses of the British Fauna (New Series) No. 15.* Published for the Linnaean Society of London and the Estuarine and Coastal Sciences Association by the Field Studies Council, Preston Montford, Shrewsbury.

*Smith, S. M, 1974 Key to the British Marine Gastropoda. *Royal Scottish Museum; Information Series; Natural History No. 2.* A useful key in pamphlet form.

*Southward, E. C. and Campbell, A. C. 2005 *Echinoderms. Synopses of the British Fauna (New Series) No. 56.* Published for the Linnaean Society of London and the Estuarine and Coastal Sciences Association by the Field Studies Council, Preston Montford, Shrewsbury.

*Stace, C. 1977 *New Flora of the British Isles.* Cambridge University Press.

*Stephenson, T. A. 1928 and 1935 *The British Sea Anemones.* 2 vols. Ray Society, London. Much more up-to-date than Gosse 1860, and beautifully illustrated. It does not include the corals.

Stephenson, T. A. and Stephenson, A. 1949 The universal features of zonation between tide marks on rocky coasts. *Journal of Ecology, 38,* 289–305. See introduction for details.

*Tattersall, W. M. and Tattersall, O. S. 1951 *The British Mysidacea.* Ray Society, London. An excellent account of the mysids with many line drawings.

*Tebble, N. 1976 *British Bivalve Seashells* (second edition). British Museum (Natural History) Publications, London. An excellent handbook for identification.

*Thompson, T. E. 1976 *Biology of Opisthobranch Molluscs* vol. 1. Ray Society, London. A very comprehensive account of the group.

*Thompson, T. E. 1978 *Biology of Opisthobranch Molluscs* vol. 2. Ray Society, London.

*Thompson, T. E. and Brown, G. H. 1976 *British Opisthobranch Molluscs. Synopses of the British Fauna (New Series) No. 8.* Published for the Linnaean Society of London and the Estuarine and Coastal Sciences Association by the Field Studies Council, Preston Montford, Shrewsbury.

Wallentinus, I. 1972 Makroskopiska alger och vattenlevande fanerogamer vid svenska osterskokusten *Zoologisk Revy;* 34, 69–84. An account of the larger marine plants of the Baltic. Text in Swedish.

*Wheeler, A. 1969 *The Fishes of the British Isles and North West Europe.* MacMillan, London. An excellent reference book on fishes.

Yonge, C. M. 1949 *The Sea Shore* New Naturalist series, Collins, London. Provides an excellent account of life on the shore. Well illustrated. Published by Fontana as a paperback in 1971.

## Websites

Encyclopedia of Marine Life of Britain and Ireland
http://www.habitas.org.uk/marinelife
Marine Conservation Society http://www.mcsuk.org/
Marine Life Information Network for Britain and Ireland http://www.marlin.ac.uk/
Wild Life of the Channel Islands http://suedalyproductions.com/index.htm

# GLOSSARY

**Aboral** describes the surface of the body opposite that which bears the mouth.

**Ambulacrum** (of echinoderms) usually a groove, with a row of tube-feet on either side; generally five per animal.

**Antenna** usually a long, slender, sensory appendage on the heads of some arthropods and some annelids.

**Asymmetrical** without symmetry, being irregular or unequal; used to describe the growth form of some animals, e.g. certain sponges.

**Benthic** dwelling in or on the seabed

**Bilateral symmetry** symmetry of an organism (e.g. a fish) which can be divided into two equal and complementary left and right halves, but which has dissimilar front and hind ends.

**Brackish** describes water usually containing less, but occasionally more, salt than is usually found in the sea.

**Byssus** hair-like filaments which attach some bivalves to rocks or plants.

**Calcareous** being made of calcium carbonate or chalk.

**Cell** smallest functional unit of a plant or animal, consisting of a nucleus surrounded by cytoplasm and bounded by a membrane, and sometimes a cell wall.

**Cephalothorax** region combining the head and thoracic segments of advanced crustaceans.

**Chaeta** bristle of polychaetes.

**Chela** leg of crustaceans which bears pincers or nippers

**Chemoreceptor** sense organ for detecting chemical stimuli as in smell or taste.

**Chitin** organic constituent of cuticle, as found in arthropods.

**Chitinous** made of chitin.

**Chordate** animal with at least a simple form of backbone (the notochord) at some stage in the life cycle; includes the vertebrates.

**Cilia** minute, filamentous structures which, by beating, may create a current and provide locomotion; visible only under the high power of a microscope.

**Cirrus** small, tentacular or sometimes finger-like appendage found in certain arthropods and polychaetes.

**Class** major subdivision of a phylum.

**Coelom** fluid-filled cavity, formed within the middle cell layer of animals.

**Commensal** organism of one species which lives in close association with one or more different species.

**Crenulate** having the edge cut into very small scallops.

**Cuticle** exterior skeleton of chitin and protein; may be tanned as in insects.

**Detritus** particles of decaying organisms accumulating, for example, on the seabed; forms the food of many invertebrate animals.

**Disc** (of an anemone) either the mouth disc which bears the tentacles, or the basal, sucker-like attachment disc; (of an ophiuroid) body excluding the arms.

**Dorsal** upper side of a bilaterally symmetrical animal (c.f. Ventral)

**Ectoparasite** parasite living on the outer surface of another organism.

**Epiphyte** plant which grows on the outer surface of another organism.

**Epizoic** describes an animal which grows on the outer surface of another organism.

**Eulittoral zone** biologically defined zone on the seashore whose uppermost limit is marked by the highest point at which barnacles occur, and whose lowermost limit is marked by the highest point at which ammarians occur (see Lewis, J. R. 1964); equivalent to the term *middle shore* as used in this book.

**Evert** turn inside out; often applied to the process of extending the proboscis of worms.

**Exhalent** breathing out; applied to respiratory streams of water in organisms or the anatomical structures by which they are conveyed.

**Foot** (of molluscs) organ on the underside of the body used in gastropods for creeping, and in bivalves for various functions including secretion of byssus, digging and burrowing.

**Free tooth** tooth not attached to jaws; as teeth on the proboscis of polychaetes such as *Nereis*.

**Free living** living unattached to any other structure.

**Frond** (of alga) all of the plant except the holdfast.

**Gamete** sperm or egg.

**Gametophyte** (of plants) generation which produces sperms and eggs.

**Genus** group of related species; many genera may form one order.

**Growth line** recognizable line or mark on a shell which indicates the start or end of a period of shell growth.

**Hermaphrodite** organism which has reproductive organs of both sexes and thus produces sperms and eggs.

**Heteromorphic** (of plants) condition where the gametophyte and sporophyte generations are dissimilar inform (c.f. Isomorphic).

**Holdfast** attachment organ of seaweeds.

**Inhalant** breathing in; applied to respiratory streams of water in organisms or the anatomical structures by which they are conveyed.

**Invertebrate** without a backbone.

**Isomorphic** (of plants) condition where the gametophyte and sporophyte generations are similar in form (c.f. Heteromorphic).

**Lamella** thin, plate-like structure or layer.

**Larva** developmental phase of an organism which usually does not resemble the adult or lead a way of life similar to it; a phase often associated with an entirely different manner of feeding from the adult and which provides a dispersive mechanism in many sedentary marine species; always terminates with the process of metamorphosis.

**Littoral** pertaining to the shore; a biologically defined zone on the seashore comprising of the eulittoral zone and the littoral fringe.

**Littoral fringe** a biologically defined zone on the seashore whose uppermost limit is marked by the highest point at which periwinkles of the genus *Littorina* occur, and whose lowermost limit is marked by the highest point at which barnacles occur, (see Lewis, J. R. 1964); equivalent to the term *upper shore* as used in this book.

**Lusitanian** applied to water masses and plankton originating from the Mediterranean and Atlantic region of Portugal.

**Mandible** jaw, especially as applied to the arthropods.

**Mantle** special region of the body wall, particularly of molluscs, which secretes the shell and encloses the mantle cavity.

**Medusa** the jellyfish phase in the life cycle of hydrozoan and scyphozoan cnidarians.

**Metamorphosis** the act of transformation of a larva into an adult.

**Neap tide** tide with the smallest range between high and low water.

**Nekton** (c.f. Plankton) swimming animals which are able to determine their position in the sea.

**Nematocyst** special cell of cnidarians which discharges threads to either sting or ensnare their prey.

**Nephridium** excretory organ of many invertebrate species.

**Niche** limiting resources and habitat of a species; determined by its interaction with a wide variety of biological, physical and chemical environmental factors.

**Notochord** skeletal tube running from front to back in some simple chordates; forerunner of the backbone of vertebrates.

**Ocellus** simple light receptor.

**Oral** relating to the mouth; in echinoderms that side of the body on which the mouth is situated (c.f. Aboral).

**Order** major subdivision of a class.

**Papilla** small outgrowing structure on the surface of an organism.

**Paragaster** main cavity inside a sponge.

**Parapodium** segmental, flap-like appendage of a polychaete annelid; usually bears chaetae or bristles.

**Parasitism** condition whereby one organism, the parasite, lives on or in another, its host, at the expense of the latter.

**Pelagic** inhabiting the surface waters of the sea.

**Pentannerism** five-fold symmetry found in echinoderms.

**Perisarc** thin, tubular, skeletal structure investing the outer surface of many hydroid polyps.

**Peristalsis** form of motion resulting from the interaction of circularly and longitudinally arranged muscles in organs like intestines and in whole animals like worms.

**Pharyngeal** relating to the pharynx.

**Pharynx** anterior region of the alimentary canal; it adjoins the gills in chordates.

**Phylum** major division of the animal kingdom which includes those animals thought to have a common evolutionary origin.

**Phytoplankton** planktonic plants (generally microscopic).

**Plankton** drifting organisms or swimming organisms which are not able to determine their position in the sea.

**Planula** simple larva (for instance) of cnidarians resembling a ball of cells; usually ciliated and hence able to move.

**Pneumatophore** organ of flotation containing gas, or a modified individual in a siphonophoran colony which subserves this function.

**Polymorphism** the occurrence of different forms of the same species, for instance in a life cycle (polymorphism in time), or in a colony (polymorphism in space).

**Polyp** sedentary, individual cnidarian such as *Hydra*, basically with a sac-like body opening only by the mouth which is generally surrounded by tentacles.

**Proboscis** special structure at the anterior end of some animals; in nemertines it is generally everted through the mouth but is itself not part of the alimentary canal; in polychaetes it is everted through the mouth and is part of the alimentary canal.

**Radial symmetry** symmetry of an organism (e.g. a cnidarian) in which the body parts are equally arranged around a median vertical axis which passes through the mouth; lacking definite front and rear ends and hence left and right sides.

**Radula** small, horny, tongue-like strip bearing teeth used by many molluscs for rasping food.

**Rhizoid** root-like structure.

**Rostrum** pointed projection at the extreme anterior end of the crustacean head.

**Salinity** measure of salt concentration of water.

**Segment** one of a (generally fixed) number of functional units of the body and normally bearing a pair of appendages; applied particularly to annelids and arthropods.

**Sessile** commonly meaning living attached to a structure such as a rock or a shell of another organism.

**Siphon** tube leading into or out of the bodies of invertebrates and used for conducting water currents; found especially in molluscs and sea-squirts.

**Species** reproductively isolated group of interbreeding organisms; usually defined by morphological characteristics.

**Spicule** minute fragment or crystal of skeletal material.

**Spiracle** vestigial gill slit found in fishes such as sharks.

**Splash zone** zone on the shore above the highest point to which the tides flow but which is under the influence of spray and salt.

**Spore** minute reproductive germ or particle produced by the asexual generation of plants, i.e. the sporophyte.

**Sporophyte** (of plants) that generation which produces asexually reproductive spores; alternates with the gametophyte.

**Spring tide** tide with the greatest range between high and low water.

**Stolon** root-like structure found in some animals and linking up individuals in a colony; in plants a horizontal branch which produces its own roots and subsequently a new individual.

**Sublittoral** biologically defined zone on the seashore which lies below the highest point to which laminarians grow, and only uncovered at the lowest tides and extending down from the shore to the shallow seabed; equivalent to the term *lower shore* as used in this book.

**Symbiotic** describes an organism of one species which lives in close association with one of another species and to the advantage of both.

**Telson** terminal, flap-like appendage of many crustaceans.

**Test** 'shell' of a sea-urchin or starfish which, strictly speaking, is an internal skeleton.

**Thallus** entire body of a lower plant, such as an alga or lichen.

**Theca** cup-shaped skeleton of a hydroid or coral polyp; the cup-shaped test of a feather-star (crinoid).

**Thecate** possessing a theca.

**Thixatropic** condition in which sand contains much water and becomes slushy when compressed, and so is able to flow.

**Torsion** twisting of the body; particularly applied to an event in the development of certain larvae, e.g. gastropods and holothuroids.

**Tube-foot** hydraulic appendage of echinoderms and part of the water vascular system.

**Tunic** the proteinaceous coat surrounding the bodies of sea-squirts and salps.

**Umbilicus** aperture in the central pillar of a snail shell.

**Umbo** part of the shell of a bivalve mollusc.

**Ventral** underside of a bilaterally symmetrical animal (c.f. dorsal).

**Vertebrate** animal with a backbone made up of vertebrae.

**Viscera** organs inside the body cavity, especially intestines, heart, liver, etc.

**Visceral** hump part of the gastropod body where most of the internal organs are housed.

**Water vascular system** hydraulic system unique to the echinoderms comprising a system of vessels and organs like the tube-feet, and fulfilling various functions, especially locomotion.

**Zoecium** casing surrounding an individual ectoproct zooid.

**Zonation** separation of plants and animals into discrete zones or communities on the shore related to the tidal levels.

**Zooid** individual animal in a colony; usually applied to the Cnidaria and Ectoprocta.

**Zooplankton** animals of the plankton.

# ACKNOWLEDGEMENTS AND INDEX

## ACKNOWLEDGEMENTS

The author and artist wish to acknowledge the great help given by so many friends and colleagues. Special thanks must be given to the following for criticism, advice and loan of material: Dr G. Boalch, Dr Bob Earll, Dr J. Fraser, the late Professor M. Godward, Professor A. G. Hildrew, Dr Sue Hiscock, Dr P. Holigan, Mrs L. M. Irvine, the late Professor M. S. Laverack, Dr J. R. Lewis, The Manchester Museum, the late Professor N. B. Marshall, Dr T. Norton, Professor D. Nichols, Mr P. Oliver, Mr and Mrs T. Pain, Queen Mary College, The University of London, The Royal Scottish Museum, Professor P. S. Rainbow, Professor J. S. Ryland, the late Dr N. Tebble, Mrs B. Thake, and Mr A. Wheeler.

The publishers would like to acknowledge the valuable assistance given by Dr G. W. Potts in the preparation of this book, and in the provision of reference material for a number of illustrations.

Original line drawings by James Nicholls, Linda Rogers Associates and Roger Gorringe.

Dr Andrew Campbell studied zoology at St Andrews and Oxford Universities, and is now senior lecturer at Queen Mary College, University of London. He has studied extensively the marine life of the European area, and has published many scientific papers on invertebrates and marine biology.

## INDEX